DATA SCIENCE AND COMPLEX NETWORKS

Data Science and Complex Networks

Real Cases Studies with Python

Guido Caldarelli

IMT, Piazza S. Francesco 19, Lucca, Italy;
ISC, Istituto dei Sistemi Complessi, CNR, Italy;
LIMS, London Institute of Mathematical Sciences, London, UK;
Linkalab, Complex Systems Computational Laboratory, Cagliari, Italy;
Catchy, Laboratory of Big Data Analysis, Roma, Italy

Alessandro Chessa

IMT, Piazza S. Francesco 19, Lucca, Italy;
ISC, Istituto dei Sistemi Complessi, CNR, Italy;
Linkalab, Complex Systems Computational Laboratory, Cagliari, Italy;
Catchy, Laboratory of Big Data Analysis, Roma, Italy

Great Clarendon Street, Oxford, OX2 6DP,
United Kingdom

Oxford University Press is a department of the University of Oxford.
It furthers the University's objective of excellence in research, scholarship,
and education by publishing worldwide. Oxford is a registered trade mark of
Oxford University Press in the UK and in certain other countries

First Edition published in 2016

Reprinted 2017 (twice)

Impression: 3

Published in the United States of America by Oxford University Press
198 Madison Avenue, New York, NY 10016, United States of America

British Library Cataloguing in Publication Data
Data available

Library of Congress Control Number: 2016949583

ISBN 978–0–19–963960–1

Printed and bound by
CPI Group (UK) Ltd, Croydon, CR0 4YY

Preface

The audience for this book comprises researchers, not necessarily trained in mathematics and physics, who have the problem of analysing and sorting out huge quantities of data. The theory of complex networks has already proved to be a good choice when looking for solutions to such problems. The reasons for that success rely on the fact that in a large variety of situations it is very easy to identify the basic units of the system under consideration; network theory models them by vertices and describes their interactions by means of links, thereby forming a clear mathematical set up for many scientific questions. A paramount example is provided by technological networks. Sites and hyperlinks are immediately described by vertices and edges. The same association for server and cables is almost as immediate. On the one hand, such a simple model reveals the inner structure of what we are observing (if we want to measure the reliability of the Internet, we need to know its skeleton). On the other hand, even if the structure we observe has no obvious network structure, we can still use network theory to explore it. An example is given by the big-data world in which we now live. Such data may span between personal information as registered by mobile phones, GPS devices, financial transactions, or results from high-throughput biological experiments. In all these cases networks can provide useful metrics to filter information. Networks allow us to filter information by selecting more relevant and important correlations and they allow for a quantitative description of any approximation made.

Exactly in this spirit, we present in this book the latest results in the field of complex networks. They can solve immediate problems when studying the Internet and the WWW, and they can help to sort information on a variety of other systems. This book is structured in such a way as to start with a specific problem and then present the theoretical tools needed to model or to sort out the most relevant information. At the end of every chapter we present the structure of the computer program needed to apply these ideas to the data.

Working copies of the software are available from the web site of the book, **http://book.complexnetworks.net**.

Practitioners (and more generally all the readers) are not required to have expertise in statistical physics or in computer programming; this text can provide all the necessary information. We are indebted to many friends and colleagues for their critical reading and suggestions, amongst them we want to cite Bruno Gonçalves and Nicola Perra; a special thank goes to Sebastiano Vigna for his precious comments.

Guido Caldarelli acknowledges for support the EU projects Multiplex nr. 317532, Simpol nr. 610704, Dolfins nr. 640772 SoBigData 654024 CoeGSS 676547.

Alessandro Chessa acknowledges support from PNR national project CRISIS-Lab (IMT School for Advanced Studies).

Contents

Introduction

Complex networks, thanks to their intuitive and interdisciplinary basis, are becoming more and more popular among scientists and practitioners (Albert and Barabási, 2002; Catanzaro and Caldarelli, 2012). Not only is basic modelling relatively simple to understand, but also the "local" network properties (i.e. relative to single vertices/nodes), such as the degree or the clustering, can immediately be visualised and computed. In fact, while the first approach may produce immediate results, less trivial quantities and measurements require a greater technical knowledge. In particular, going from local graph quantities to more global ones (such as the resilience or the optimisation) and passing through two- and three-vertices properties (such as reciprocity and assortativity) may prove to be less intuitive. In parallel with the success of network theory, we are also witnessing another phenomenon, that is, the growth in numbers and size of data repositories. Every day an increasing flow of news, media, comments in the form of blogs, likes, and sharing on social networks is produced and stored. Algorithms and techniques, some simple and some more elaborate, are necessary to extract valuable information from this amount of data.

The aim of this book is to show how network theory is a crucial tool both to model and to filter information in these data-producing systems. In this spirit we want to guide readers through this fascinating field by presenting both datasets and network concepts. The idea is to start from basic (and indeed simple and immediate) concepts, moving towards more elaborate quantities and to give the instruments (especially the software) to the reader to continue their research even further. Towards this purpose, following the recent success of complex networks, there is now a series of books and manuals on the topic. Spanning between introductory and more advanced courses (Dorogovtsev and Mendes, 2003; Caldarelli, 2007; Barrat *et al.*, 2008; Dorogovtsev, 2010; Newman, 2010) and considering the various reviews and publications available (Newman, 2003; Barabási and Bonabeau, 2003; Boccaletti *et al.*, 2006; Estrada *et al.*, 2010), there is now an important corpus of work available to scientists and practitioners. With respect to all of these books, we want to present a different approach where the activity starts by working directly on raw data. Indeed, the concepts written in theoretical books, need to be used in real situations, i.e. in most of the cases they need to be translated into software. Since in practical situations the data we are interested in are often in a raw and unformatted form (very different to the exempla we see in textbooks and publications), the absence of powerful pieces of code results in an obstacle to the work of researchers. This situation is rather general, since the need for fast and easy implementation of the basic concepts of complex networks is growing in many different fields, namely biology, ecology, finance and economics, not to mention the interest for policy makers, stakeholders, and practitioners. In this book we have selected a series of topics and we have shown the data available for them.

Through the analysis of such "physical problems" we introduce both the theoretical concepts of network theory and their coding in Python. Since many softwares have already been developed to such a purpose, we made a choice to consider one of the most successful (at least at present), i.e. the library Networkx. The basic concepts of networks as degree, degree sequence, clustering, community detection, reciprocity, centrality, and/or betweenness are then introduced chapter by chapter, with specific examples and with their related codes.

It has not escaped our notice that such an approach is not only a different way to present well-known results, but it rather reflects the present scientific situation of data deluge. The choice to start with data and then introduce theory is not only a way to understand the world in which we are living, but also good practice for trained scientists. This choice also helps in addressing complex networks as an important tool to mine technological, social, and financial data. Instruments such as minimal spanning trees (or forest) and maximal planar graphs filter information in such a way as to represent the most important features. This will be particularly evident when considering financial networks, where minimal spanning tree techniques are particularly useful in clustering stocks from similar sectors. Since science is also interested in the reproduction of results and not only in their description, we have added a final part on modelling. Indeed, several models have been presented so far in the field of complex networks. By using the same datasets and topics described in this book, we have the opportunity of introducing all of them.

The plan of the work as outlined starts from definition of local quantities and proceeds towards more general concepts of community structure and modelling.

- In Chapter 1 we start with food webs, to show how to represent a graph by means of an adjacency matrix, and we introduce concepts related to the property of a single vertex and its neighbours such as the *degree* and *clustering degree sequence*.
- In Chapter 2 we analyse the World Trade Web under various aggregations to move to two-vertices properties such as *assortativity* and *reciprocity*, and we also introduce a special kind of graph that is particularly useful for the analysis of social networks, that is *the bipartite graph*.
- in Chapter 3 we study the structure of the Internet and we introduce various measures of centrality used to assess the *robustness* of the structure.
- in Chapter 4 we study WWW and Wikipedia, introducing another measure of centrality given by *PageRank*. We then study Wikipedia as a playing field for the analysis of *communities* in particular we introduce the *Girvan–Newman* community-detection method based on betweenness.
- In Chapter 5 we take into consideration some instances of financial networks. Analysis of this dataset allows us to define *centrality* measures as the *DebtRank*.
- in Chapter 6 we present a series of models among which are the *random graph model*, the *configuration model*, the *gravity model*, the *fitness model*, the *Barabási–Albert model*, the *copying model*, the *dynamic fitness models*, and a methodology to reconstruct networks from missing information.

What you need to start coding is to download and install Python 2.7.* (in Linux and Mac is in the operating system by default) and possibly the Anaconda package that contains all the scientific libraries for the code presented in this book (https://www.continuum.io/downloads). We shall use almost everywhere the libraries Numpy, Scipy, Matplotlib and NetworkX, easily retrievable from the web and included in Anaconda. Specific dependencies are for the libraries

- `BeautifulSoup` (https://www.crummy.com/software/BeautifulSoup) in Chap. 3;
- `Twython` (https://twython.readthedocs.io) in Chap. 4;
- `yahoo_finance` (https://pypi.python.org/pypi/yahoo-finance) in Chap. 5;
- `sympy` (http://www.sympy.org/) in Chap. 6.

In every chapter you will find the codes typed within a box. Anyhow it is not necessary to copy it into your terminal. We set up the web site http://book.complexnetworks.net where you can download all of them in the form of a Jupyter Notebooks (http://jupyter.org/) with the data associated to get the results showed in the book. In the same place you will find all the instructions to properly install the needed software. We put every effort in checking the text for typos and code bugs, in the unlikely event that some of them escaped our notice, please refer to the web site where updated versions will be published as soon as they will be available.

Lucca, April 2016

Guido Caldarelli & Alessandro Chessa

1
Food Webs

1.1 Introduction

Food webs are collections of predation relationships between species living in the same habitat. As shown in Fig.1.1 such collections appear (and actually are) quite difficult to visualise and any reductionist approach misses the point of functionality description (not to mention the prediction of any future behaviour). The necessity of a comprehensive approach makes food webs a paramount example of a complex system (Havens, 1992; Solé and Montoya, 2001; Montoya and Solé, 2002; Stouffer *et al.*, 2005). Already in 1991, food webs had been described as a community of predators and parasites as "complex, but not hopelessly so" (Pimm *et al.*, 1991). More specifically, following a traditional approach (Cohen, 1977), ecologists distinguish among

- *community webs* defined by picking in the same habitat (or set of habitats) a group of species connected by their predation relationships;
- *sink webs* made by collecting all the prey eaten by one or more predators and recursively the prey of these prey;
- *source webs* made by collecting all the predators of one or more species and recursively the predators of such predators.

In any of these webs, prey or predators do not necessarily correspond to distinct species. Rather, the same species can appear more than once in different roles (i.e. different stages in the life cycle of an organism). Also, since cannibalism is present (both at the same and at different stages of growth of individuals) a single species can be both prey and predator.

Two problems arise: firstly, by considering the definition of community webs, it is clear that working out who eats whom can be in principle rather complicated and typically many years of field observation are necessary to spot particularly rare events. Secondly, when we list species in a given habitat (Willis and Yule, 1922) we are more inclined to spot very small differences between large animals (difference in colours of eagles or tigers), while we do not notice the differences between species of similar bacteria. As regards the first issue, there is no other solution than working hard to improve the quality of the data collected. Indeed, we now understand that scarcity of observation can lead to partial knowledge of predations, thereby underestimating the role of certain species in the environment (Martinez, 1991; Martinez *et al.*, 1999). For the second issue it is customary to reduce the observational bias by introducing the concept of *trophic species* (Memmott *et al.*, 2000). Those are a coarse grained version of species and they can be obtained by lumping together different organisms when they feed on the same prey and they are eaten by the same predators.

Data Science and Complex Networks. First Edition. Guido Caldarelli and Alessandro Chessa.

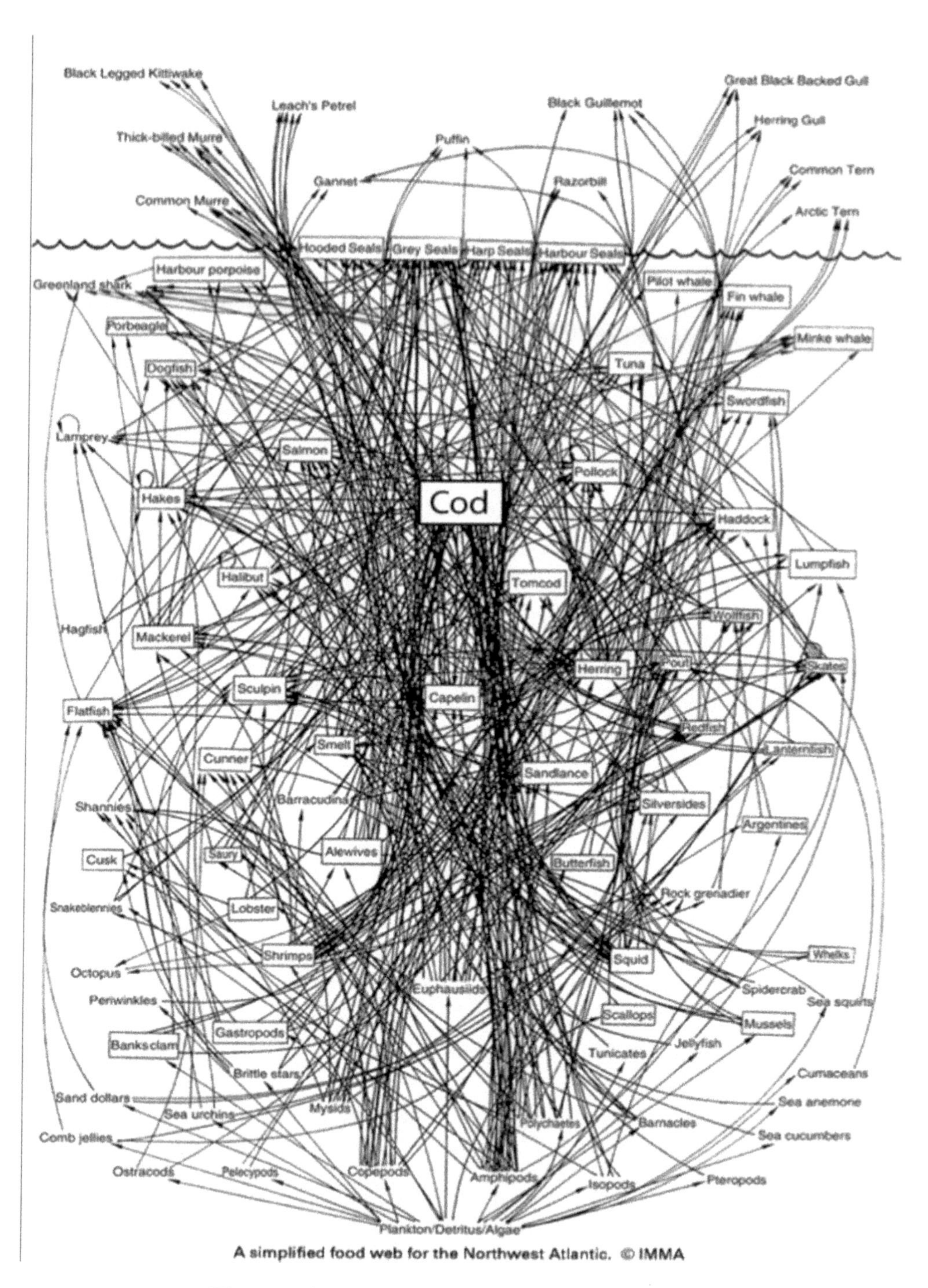

A simplified food web for the Northwest Atlantic. © IMMA

Fig. 1.1 An example of a partial marine food web.

Food webs are a typical example of a system where we find a natural characterisation in the form of a network. Directed edges (prey $\rightarrow$ predators) connect the various vertices (trophic species). Not surprisingly then, many ecological quantities have their counterpart in graph theory:

- number of trophic species S that are the number n of vertices (measure of a graph);
- number of links L that are the number m of edges (size of a graph);
- number of possible predations $\simeq S^2$ (with cannibalism $= S^2$, without $= S(S-1)$) corresponding to the size of a complete directed graph;
- connectance $C \simeq L/S^2$ of a given habitat, that is, the ratio between the number of edges present with respect to those that are possible (see previously);
- number of prey per species, that is the in-degree of the vertex and similarly the number of predators per species, that is the out-degree of the vertex;
- number of triangulations between species (Huxham *et al.*, 1996). In a directed graph it is related to the motifs structure, in a simplified form of an undirected graph with clustering.

Food webs, in particular, also show some particular behaviour that is rather uncommon in other fields: vertices can be naturally ordered according to a scale of levels (see Fig. 1.2). The vertices in the first level are species "predating" only water, minerals, and sunlight energy. In ecology those are known as "basal species" or primary producers. Species in the second level are those who predate on the first level (irrespective of the fact that they can also predate other species). This concept can be iterated and in general the *level* of one species is related to the minimum path to "basal species". In this way, we can cluster together trophic species into three simple classes:

- *basal species* B that have only predators;
- *top species* T that have only prey;
- *intermediate species* I that have both.

This allows us to define several quantities that we can use to describe the various food webs:

- Firstly, the proportion of such classes (Pimm *et al.*, 1991);
- Secondly, the proportion of the different links between the classes (BI, BT, II, IT);
- Finally, the ratio prey/predators (Cohen, 1977) $= (\#B + \#I)/(\#I + \#T)$.

We remind that codes, data and/or links for this chapter are available from **http://book.complexnetworks.net**

1.2 Data from EcoWeb and foodweb.org

A traditional source of data for species is given by the machine readable dataset EcoWeb,[1] presenting 181 small food webs. In all of them the number of species is

[1] https://dspace.rockefeller.edu/handle/10209/306

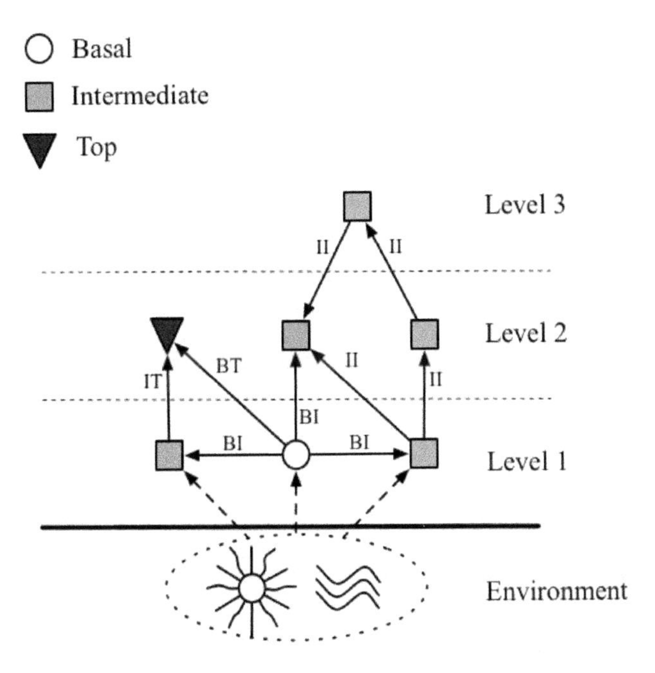

Fig. 1.2 Structure of a food web with the distinction in levels and classes of species.

rather small, even if in more recent datasets there is a somewhat larger number of species.

- St Martin Island (Goldwasser and Roughgarden, 1993) with 42 trophic species;
- St Marks Seagrass (Christian and Luczkovich, 1999) with 46 trophic species;
- Another grassland (Martinez *et al.*, 1999) with 63 trophic species;
- Silwood Park (Memmott *et al.*, 2000) with 81 trophic species;
- Ythan Estuary (without parasites) (Hall and Raffaelli, 1991) with 81 trophic species;
- Little Rock Lake web (Martinez, 1991) with 93 trophic species;
- Ythan Estuary (with parasites) (Huxham *et al.*, 1996) with 126 trophic species.

We have listed these food webs in the material attached to this book. Other datasets can be downloaded from:

- `http://vlado.fmf.uni-lj.si/pub/networks/data/bio/foodweb/foodweb.htm`
- `http://datadryad.org/resource/doi:10.5061/dryad.b8r5c`
- `https://networkdata.ics.uci.edu/`
- `http://globalwebdb.com/`

A good public resource where it is possible to download publications, software tools, and data is the basic PEaCE Lab web site at http://foodwebs.org.

Fig. 1.3 A simple graph and its adjacency matrix.

1.3 Store and measure a graph: size, measure, and degree

1.3.1 The adjacency matrix

Graphs can be drawn (note that this can be done in many ways) either as a picture or by means of a mathematical representation known as an "adjacency matrix", as shown in Fig.1.3. While the picture is useful for visualising immediately some properties of the graph (typically local ones), the matricial form is very useful for computing others (typically global). The adjacency matrix A is a square table of numbers ($n \times n$, where n is the number of vertices in the graph), where the elements in row i and column j are equal to 1 if we have an edge between vertex i and j, and equal to 0 otherwise. Such structures can be stored by computers and this representation turns out to be particularly useful when dealing with large datasets. It would be insufficient to believe that matrices are only useful as computer representations of graphs. Spectral properties (related to the matricial form) of graphs can reveal many properties, starting with community structure. We shall see these points in detail later in this book.

The simplest case of an undirected graph gives symmetric matrices A. If there is an edge between i and j there is also an edge in the opposite direction joining j and i, so that $a_{ij} = a_{ji}$. Instead, if there is a direction in the linkage we have an arc starting from the node i and pointing to the node j, but the opposite is not necessarily the case, in this way $a_{ij} \neq a_{ji}$ and the matrix is no longer symmetric. Undirected networks are the simplest form of graph we can deal with. Symmetric matrices indicate that we have only one kind of link connecting two vertices. In the case of the Internet this means that a cable connecting two servers can be used both for sending part of a mail from the first to the second and also for answering the mail from the second to the first. Asymmetric matrices describe situations where the path cannot be followed in reverse. If you put a link from your web page to the web page of your favourite football team or Hollywood actor, you will not necessarily (and probably seldom) receive a link back to your page. Oriented networks arise in a variety of situations. On the web in social networks where we have e-mail, "likes" on Facebook posts, and retweets on Twitter that are not symmetric; in finance where borrowing money is not lending it; in economics where we have trade between nations, and finally in biology where the predation relations in food webs are typically not reciprocal.

In this section we present the basic quantities used to describe a graph (some have already been mentioned), together with an example of the Python code.

Read the adjacency matrix

Starting from this formulation it is relatively easy to cast it in Python code. We can either start with the matrix representation of the graph or with the direct representation of nodes and edges from the Networkx library. In the first case, we will first represent the graph through the adjacency matrix using the basic list data type (in this case list of lists) in the following way:

```
adjacency_matrix=[
                  [0,1,0,1],
                  [1,0,1,1],
                  [0,1,0,0],
                  [1,1,0,0]
                  ]
```

The basic Python statement for iterative cycles is a little bit different from the usual programming languages, like Fortran, C/C++, and the like. The iteration is supposed to run over a list and, for example, in a simple case of an index i running from 1 to 5 the syntax will be:

```
for i in [1,2,3,4,5]:
    print i

#OUTPUT
1
2
3
4
5
```

We can browse the matrix rows using the "for" Python statement:

```
for row in adjacency_matrix:
    print row

#OUTPUT
[0, 1, 0, 1]
[1, 0, 1, 1]
[0, 1, 0, 0]
[1, 1, 0, 0]
```

Note the indentation of the "print" statement, that is mandatory in Python, even if its length is not fixed.

To get each single matrix element we will nest another for cycling to extract each element of the rows:

```
for row in adjacency_matrix:
    for a_ij in row:
        print a_ij,
```

```
    print "\r"

#OUTPUT
0 1 0 1
1 0 1 1
0 1 0 0
1 1 0 0
```

The comma prevents the new line adding simply a space in the visualisation of the row, while the special character "\r" stands for a carriage return.

In the case of directed networks the adjacency matrix is not symmetric, like for a food web; if a non-zero element is present in row 2, column 3, this means there is an arc (directed edge) from node 2 towards node 3:

```
adjacency_matrix_directed=[
                     [0,1,0,1],
                     [0,0,1,0],
                     [0,0,0,1],
                     [0,0,0,0]
                     ]
```

1.3.2 Size, measure, connectance

The simplest scalar quantities defined in the title can be computed easily in the case of the various food webs. We recall them here:

- the number of species S, that in graph theory corresponds to the number of vertices n which is the *measure* of the graph;
- the number of predations L, that in graph theory corresponds to the number of edges m, which is the *size* of the graph;
- since in this case we can distinguish among the different nature of vertices (i.e. B,I,T), we can measure proportions of species and links between them;
- The connectance $C \simeq L/S^2$, corresponding to the density of the graph (actual edges present divided by the maximum possible number).

Basic statistics

```
#the number of species is the number of rows or columns of
#the adjacency matrix
num_species=len(adjacency_matrix_directed[0])

#the number of links or predations is the non zero elements
#of the adjacency matrix (this holds for directed graphs
```

```
num_predations=0
for i in range(num_species):
    for j in range(num_species):
        if adjacency_matrix_directed[i][j]!=0:
            num_predations=num_predations+1

#to check if a specie is a Basal (B), an Intermediate (I) or
#a Top (T) one  we have to check the presence of 1s both in
#the row and in the column of each specie
row_count=[0,0,0,0]
column_count=[0,0,0,0]
for i in range(num_species):
    for j in range(num_species):
        row_count[i]=row_count[i]+adjacency_matrix_directed[i][j]
        column_count[j]=column_count[j]+ \
        adjacency_matrix_directed[i][j]

number_B=0
number_I=0
number_T=0

for n in range(num_species):
    if row_count[n]==0:
        number_T+=1
        continue
    if column_count[n]==0:
        number_B+=1
        continue
    else:
        number_I+=1

print "number of species", num_species
print "number of predations", num_predations
print "classes Basal, Top, Intermediate: ",number_B,number_T,number_I
print "connectance", float(num_predations)/float(num_species**2)

#OUTPUT
number of species 4
number of predations 4
classes Basal, Top, Intermediate:  1 1 2
connectance 0.25
```

1.3.3 The degree

The simplest quantity that characterises the vertex is the number of its connections. This quantity is called the *degree*, sometimes (mostly by physicists) called "connectivity". The degree of a vertex indicates the connections of this vertex; the degree is thereby a "local" quantity (you need to inspect only one vertex to find its degree). In the following we shall see non-local measures of graphs, which involve two or more vertex neighbours, and also measures that are "global" and need an inspection of the whole system to be computed (i.e. betweenness). The frequency distribution of this quantity is traditionally used as a signature of "complexity", in the sense that the presence of long tails (or a scale-free distribution) is interpreted as a signature of long-range correlation in the system. When the graph is oriented we can distinguish between in-degree and out-degree. The former accounts for ingoing links (for example the energy we receive when eating another living organism), the latter accounts for outgoing links (as for example the hyperlinks we put on our web page to other pages we like). Once we have the matrix of the graph, the degree can easily be computed. If we want to know the degree k_i of the vertex i we simply sum the various elements a_{ij} on the various columns j, i.e.

$$k_i = \sum_{j=1,n} a_{ij}. \tag{1.1}$$

If the graph is oriented, the sum along the rows of the (non-symmetric) matrix A is different from the sum along the columns (that is not the case when the matrix is symmetric and the graph is not oriented). In one case we get the out-degree k_i^O of node i, while in the opposite we get the in-degree k_i^I of node i. In formulas,

$$k_i^I = \sum_{j=1,n} a_{ij}, \qquad k_i^O = \sum_{j=1,n} a_{ji}. \tag{1.2}$$

When the graph is weighted we can extend the previous definition, by distinguishing between the number of connections (degree) and the weighted degree or *strength* s , that is the sum of the relative weights of those links. Also in this case, we can use the matrix representation. Now every element a_{ij}^w takes the value of the weight between i and j, we have

$$s_i = \sum_{j=1,n} a_{ij}^w. \tag{1.3}$$

Typically in real situations there is a power-law relation between strength and degree (Barrat *et al.*, 2004)

Degree

With this matrix representation we can calculate the degree for a specific node (in this case the node "2"):

```
#for the undirected network
degree_node_2=0
for j in adjacency_matrix[1]:
```

```
    degree_node_2=degree_node_2+j
print "degree of node 2:",degree_node_2

#and for the directed case we already calculated the sum over
#the rows and columns for the adjacency_matrix_directed
out_degree_node_3=row_count[2]
in_degree_node_4=column_count[3]

print "out_degree node 3:",out_degree_node_3
print "in_degree node 4:",in_degree_node_4

#OUTPUT
degree of node 2: 3
out_degree node 3: 1
in_degree node 4: 2
```

Remember that the indices in Python data structures start from “0” and so the row “2” is marked as “1”.

Degree in Networkx

The equivalent procedure in Networkx will be:

```
import networkx as nx

#generate an empty graph
G=nx.Graph()

#define the nodes
G.add_node(1)
G.add_node(2)
G.add_node(3)
G.add_node(4)

#link the nodes
G.add_edge(1,2)
G.add_edge(1,4)
G.add_edge(2,3)
G.add_edge(2,4)

#degree of the node 2
print G.degree(2)

#OUTPUT
3
```

1.4 Degree sequence

When dealing with large networks we need only to coarse grain the information on connections by giving the degree sequence, that is the list of the various degrees in the graph. Such information can be summarised by making a histogram of the degree sequence (a typical and traditional statistical analysis done for complex networks that serves as a benchmark to describe the suitability of various models of network growth). Please note that while we can associate a degree sequence to any graph, obviously not all sequences of numbers can produce a graph (see also Sec. 6.4). For example, the sum of all the degrees in an undirected graph must be an even number (we are counting every edge twice, then the sum of the elements in the degree sequence gives $2E$, where E is the total number of edges). As a consequence, any degree sequence whose sum is odd, cannot form a graph. Furthermore even if the sum of elements in the degree sequence is even, most configurations are impossible (imagine a degree larger than the number of vertices present). Empirically, whenever graphs are made from a large number of vertices, it becomes more and more difficult to check if a given degree sequence is actually describing a graph or not.

We shall see more on these topics in Chapter 6, for the moment let us focus only on the passage from the graph to the degree sequence. A simple way to obtain the degree sequence starting from the previous Python formulation is to generalise the code in order to compute the degree for each row, as follows:

Degree sequence

```
degree_sequence=[]
for row in range(len(adjacency_matrix)):
    degree=0
    for j in adjacency_matrix[row]:
        degree=degree+j
    degree_sequence.append(degree)

print degree_sequence
```

and the output will be:

```
#OUTPUT
[2, 3, 1, 2]
```

1.4.1 Plotting the degree sequence, histograms

As mentioned previously, when the network is large, we want a single plot or image that might help us in describing the graph. A histogram is the best choice for that purpose and it is important to learn how to draw these objects from analysis of the raw data. In practice, we must count how many times we have a vertex whose degree is $1, 2$, etc. This number is plotted against the degree values as in Fig. 1.4.

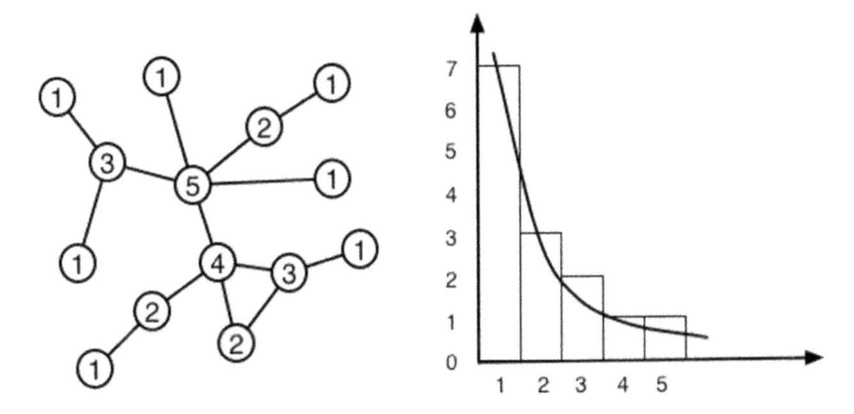

Fig. 1.4 The way in which the degree sequence is computed and plotted by means of a histogram. Node labels are the degrees.

Histogram

In Python it is extremely easy to plot any kind of graphs and one of the most popular libraries is Matplotlib. In order to get the histogram of the previous degree sequence we simply issue:

```
import matplotlib.pyplot as plt

plt.hist([1,1,1,1,1,1,1,2,2,2,3,3,4,5],bins=5)
plt.show()
```

1.5 Clustering coefficient and motifs

The *clustering coefficient* is a standard, basic measure of the community structure at local scale. Imagine a network of friendship (visualised as edges) between persons (vertices). The clustering coefficient gives the probability that if Frank is a friend of John and Charlie, also John and Charlie are friends with each other. For graphs this means that if we focus on a specific vertex i connected to other vertices, the clustering coefficient c_i measures the probability that the destinations of these vertices are also joined by a link. If all the connections are equiprobable, we just count the frequency of such connections, that is, we measure the number of triangles insisting on a particular vertex, as shown in Fig. 1.5 a. Another measure used is the clustering coefficient $c(k)$ of vertices whose degree is k. This is the average of all the values of the clustering coefficients made with all the vertices whose degree is k,

$$c(k) = \frac{\sum_{i=1,N} c_i \delta_{k_i,k}}{N_k}, \tag{1.4}$$

where N_k is the number of vertices whose degree is k and $\delta_{k_i,k} = 1$ if $k_i = k$ and 0 otherwise. Real networks are often characterised by a clustering larger than expected

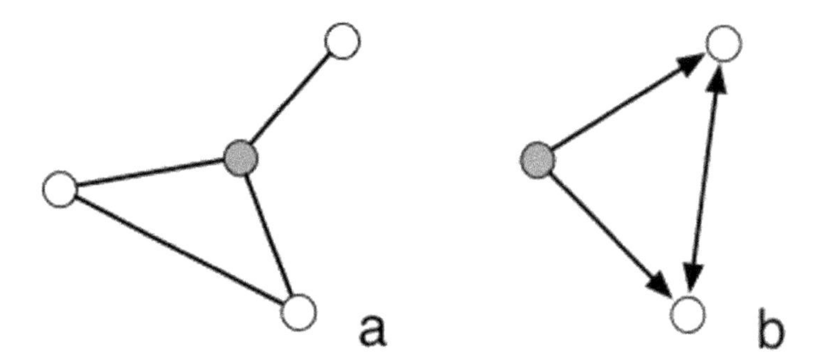

Fig. 1.5 (a) Case of a vertex (in grey) whose clustering is 1/3 (one triangle out of the possible three. (b) A feed-forward loop.

from a series of randomly connected vertices. It is worth noticing that in the case of directed networks as with food webs, it is somewhat difficult to determine which triangles must to be considered, since now the edges have a direction. By considering this further degree of freedom, a simple triangular structure can assume a variety (nine) of different shapes. Such shapes are called *motifs* and their statistics can reveal something about the density of the system at a local scale. For example, food webs are characterised by the net dominance of a "feed-forward" loop, as shown in Fig. 1.5 b. Similarly, motifs can be used to detect early signals of collapse in the particular case of financial networks (Squartini *et al.*, 2013).

Clustering coefficient

As for the node degree previously defined, we can code the clustering coefficient for a specific node. For example, looking at Fig. 1.5 and node "2" we can express the clustering coefficient computing the connections between the neighbours of node "2" and dividing by all the possible connections among them (degree*(degree-1)/2). First we compute the list indices of the neighbours of "2":

```
row=1 #stands for the node 2
node_index_count=0
node_index_list=[]
for a_ij in adjacency_matrix[row]:
    if a_ij==1:
        node_index_list.append(node_index_count)
    node_index_count=node_index_count+1
print "\r"

print node_index_list
```

and the list in the case of node "2" will be

```
#OUTPUT
[0, 2, 3]
```

then we will check all of the possible neighbour couplings for whether a link actually exists:

```
neighb_conn=0
for n1 in node_index_list:
    for n2 in node_index_list:
        if adjacency_matrix[n1][n2]==1:
            neighb_conn=neighb_conn+1

#we have indeed count them twice...
neighb_conn=neighb_conn/2.0

print neighb_conn

#OUTPUT
1.0
```

and in our case the result is simply 1. Finally the clustering coefficient for node "2" is given by the expression

```
clustering_coefficient=neighb_conn/ \
(degree_node_2*(degree_node_2-1)/2.0)

print clustering_coefficient
```

where the final result is 0.333333333333.

1.5.1 Ecological level and categories between species, bowtie

One of the distinctive features of food web data is the possibility of arranging the vertices along different levels defined by the distance from the environment (as usual, the distance in a graph corresponds to the minimum number of edges to travel between two vertices). As a result we can define categories according to the in/out links relating to the predation. All the species that have no predations are indicated as top (T), all the species with no prey (apart from the environment) are indicated as basal (B). All the others are intermediate (I). Apart from the basal species, the intermediate or top species can be more or less distant from the environment. Probably (but not necessarily!), species at the lowest level are likely to be basal, while species on the highest levels are likely to be top ones. The study of universality in number of levels and composition is one of the traditional quantitative ecological analysis in the quest for food web universality. In order to identify the various levels in the food web network we need an algorithm able to compute the distance between all pairs of nodes. A generalisation of this concept of levels and classes of nodes is given by the concept of bowtie, a structure that was first noticed in the World Wide Web (Broder *et al.*, 2000), and late in economics (Vitali *et al.*, 2011) and financial systems. In any directed network you can determine a set of nodes mutually reachable one from another. They form the strongly connected component (SCC). Those from which you arrive at SCC are the IN component. Those reachable from the SCC form the OUT component. In spite of the technical differences between top species and OUT components, the two structures have some similarities (see Fig. 1.6).

Calculating the bowtie structure for a food web network

```
#loading the network
file_name="./data/Ythan_Estuary.txt"

DG = nx.DiGraph()

in_file=open(file_name,'r')
while True:
    next_line=in_file.readline()
    if not next_line:
        break
    next_line_fields=next_line[:-2].split(' ')
    node_a=next_line_fields[1] #there is a space in the beginning
                               #of each edge
    node_b=next_line_fields[2]
    DG.add_edge(node_a, node_b)

#deleting the environment
DG.remove_node('0')

#getting the biggest strongly connected component
scc=[(len(c),c) for c in sorted( nx.strongly_connected_components \
                               (DG), key=len, reverse=True)][0][1]

#preparing the IN and OUT component
IN_component=[]
for n in scc:
    for s in DG.predecessors(n):
        if s in scc: continue
        if not s in IN_component:
            IN_component.append(s)

OUT_component=[]
for n in scc:
    for s in DG.successors(n):
        if s in scc: continue
        if not s in OUT_component:
            OUT_component.append(s)
```

```
#generating the subgraph
bowtie=list(scc)+IN_component+OUT_component
DG_bowtie = DG.subgraph(bowtie)

#defining the proper layout
pos={}
in_y=100.
pos['89']=(150.,in_y)

in_step=700.
for in_n in IN_component:
    pos[in_n]=(100.,in_y)
    in_y=in_y+in_step

out_y=100.
out_step=500.
for out_n in OUT_component:
    pos[out_n]=(200,out_y)
    out_y=out_y+out_step

pos['90']=(150.,out_y)

#plot the bowtie structure
nx.draw(DG_bowtie, pos, node_size=50)

nx.draw_networkx_nodes(DG_bowtie, pos, IN_component, \
                       node_size=100, node_color='Black')
nx.draw_networkx_nodes(DG_bowtie, pos, OUT_component, \
                       node_size=100, node_color='White')
nx.draw_networkx_nodes(DG_bowtie, pos, scc, \
                       node_size=200, node_color='Grey')

savefig('./data/bowtie.png',dpi=600)
```

The simplest algorithm to determine paths and distances is an exploration known as Breadth First Search (BFS).

Distance with Breadth First Search

As shown in Fig. 1.7 the strategy to compute the distance from the root node is to explore all the accessible neighbours not already visited.

```
#creating the graph
G=nx.Graph()
G.add_edges_from([('A','B'),('A','C'),('C','D'),('C','E'),('D','F'),
```

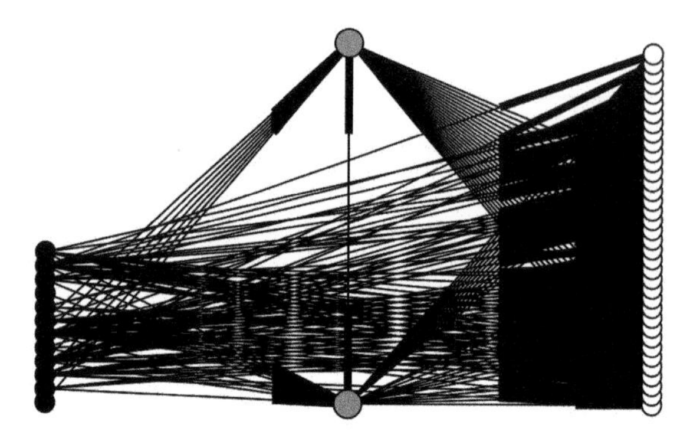

Fig. 1.6 A representation of a bowtie structure for the Ythan Estuary food web network. On the left the IN component in black. In the middle the two nodes of the strongly connected component in grey. On the right the OUT component in white.

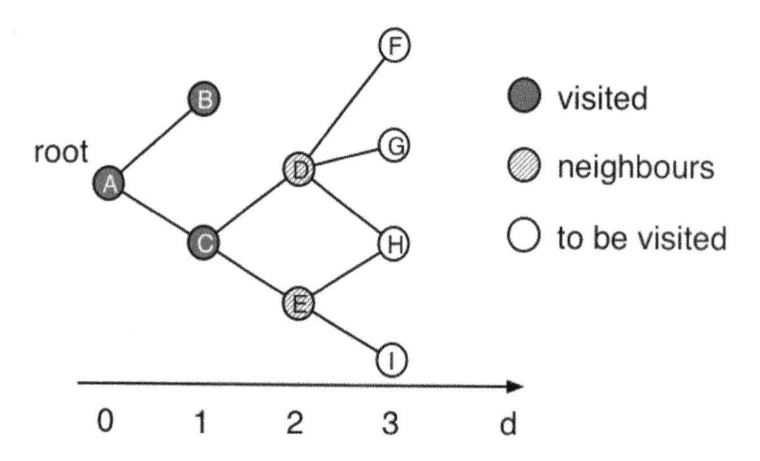

Fig. 1.7 The procedure of the Breadth First Search (BFS) algorithm. Starting from a root node at distance $d = 0$ at successive time steps we explore the neighbours at increasing distances.

```
('D','H'),('D','G'),('E','H'),('E','I')])

#printing the neighbors of the node 'A'
print G.neighbors('A')
```

```
#OUTPUT
['C', 'B']

root_node='A'
queue=[]
queue.append('A')
G.node['A']["distance"]=0
while len(queue):
    working_node=queue.pop(0)
    for n in G.neighbors(working_node):
        if len(G.node[n])==0:
            G.node[n]["distance"]=G.node[working_node]["distance"]+1
            queue.append(n)
for n in G.nodes():
    print n,G.node[n]["distance"]

#OUTPUT
A 0
C 1
B 1
E 2
D 2
G 3
F 3
I 3
H 3
```

As previously mentioned, we need trophic species to remove dimensional bias in the food webs; this is done by concentrating in a single node all the nodes with the same prey/predator pattern. In Table 1.1 we show trophic versions of the food web datasets we presented at the beginning of this chapter and their fundamental measures (Dunne *et al.*, 2002; Caldarelli *et al.*, 1998). In the following we will compute them, step by step, with simple Python code.

The first thing to do is to load the dataset in the shape of a network:

Reading the file with food web data

```
file_name="./data/Little_Rock_Lake.txt"

DG = nx.DiGraph()

in_file=open(file_name,'r')
while True:
```

```
        next_line=in_file.readline()
        if not next_line:
            break
        next_line_fields=next_line[:-2].split(' ')
        node_a=next_line_fields[1] #there is a space in the beginning
                                   #of each edge
        node_b=next_line_fields[2]
        print node_a,node_b
        DG.add_edge(node_a, node_b)

#OUTPUT
0 11
0 61
0 80
0 123
0 124
...
```

Once the specific food web has been loaded in the Networkx structure we can operate on it. The first thing to do is to generate trophic versions of this network. We will use extensively the property of the dictionary key to be a complex data structure. In the present case we will use the list/tuple as a pattern to identify the particular trophic species.

Defining the trophic pattern key

```
def get_node_key(node):
    out_list=[]
    for out_edge in DG.out_edges(node):
        out_list.append(out_edge[1])
    in_list=[]
    for in_edge in DG.in_edges(node):
        in_list.append(in_edge[0])
    out_list.sort()
    out_list.append('-')
    in_list.sort()
    out_list.extend(in_list)
    return out_list
```

Leveraging from this pattern function we can extract the trophic species through the following function:

Grouping the trophic species and regenerating the trophic network

```
def TrophicNetwork(DG):
    trophic={}
    for n in DG.nodes():
        k=tuple(get_node_key(n))
        if not trophic.has_key(k):
            trophic[k]=[]
        trophic[k].append(n)
    for specie in trophic.keys():
        if len(trophic[specie])>1:
            for n in trophic[specie][1:]:
                DG.remove_node(n)
    return DG

#deleting the environment
DG.remove_node('0')

TrophicDG=TrophicNetwork(DG)
print "S:",TrophicDG.number_of_nodes()
print "L:",TrophicDG.number_of_edges()
print "L/S:",float(TrophicDG.number_of_edges())/ \
TrophicDG.number_of_nodes()

#OUTPUT
S: 93
L: 1034
L/S: 11.1182795699
```

Categories in a food web network are in relation to in/out links of each species. We have the basal ones (B) that are the prey (outgoing links only), the top species T (ingoing links only) which are only predators, and finally the intermediate species I (with in- and outgoing links) which are both prey and predators. Here is the Python code that categorised the "Little Rock" food web network introduced before:

Classes in food webs

```
def compute_classes(DG):
    basal_species=[]
    top_species=[]
    intermediate_species=[]
    for n in DG.nodes():
```

```
        if DG.in_degree(n)==0:
            basal_species.append(n)
        elif DG.out_degree(n)==0:
            top_species.append(n)
        else:
            intermediate_species.append(n)
    return (basal_species,intermediate_species,top_species)

(B,I,T)=compute_classes(TrophicDG)
print "B:",float(len(B))/(len(B)+len(T)+len(I))
print "I:",float(len(I))/(len(B)+len(T)+len(I))
print "T:",float(len(T))/(len(B)+len(T)+len(I))

#OUTPUT
B: 0.129032258065
I: 0.860215053763
T: 0.010752688172
```

Finally, we compute the proportion of the links among the various classes previously defined and the ratio prey/predators, defined as: $(\#B + \#I)/(\#I + \#T)$.

Proportion of links among classes and ratio prey/predators

```
def InterclassLinkProportion(DG,C1,C2):
    count=0
    for n1 in C1:
        for n2 in C2:
            if DG.has_edge(n1,n2):
                count+=1
    return float(count)/DG.number_of_edges()

print "links in BT:",InterclassLinkProportion(TrophicDG,B,T)
print "links in BI:",InterclassLinkProportion(TrophicDG,B,I)
print "links in II:",InterclassLinkProportion(TrophicDG,I,I)
print "links in IT:",InterclassLinkProportion(TrophicDG,I,T)

#Ratio prey/predators
print "P/R:",float((len(B)+len(I)))/(len(I)+len(T))

#OUTPUT
links in BT: 0.000967117988395
links in BI: 0.0909090909091
```

Experimental data						
	Silwood	Grassland	St Marks	St Martin	Ythan	L. Rock
S	16	15	29	42	83	93
L	33	30	262	203	398	1034
L/S	2.0	2.0	9.0	4.8	4.8	11.1
B (%)	21	13	10	14	5	13
I (%)	49	74	90	69	59	86
T (%)	30	13	0	17	36	1
TB (%)	10	3	0	3	1	0
IB (%)	29	10	13	19	10	9
II (%)	29	57	87	53	51	91
TI (%)	32	30	0	25	38	0
P/R	0.89	1.0	1.11	0.97	0.67	1.13

Table 1.1 Some basic quantities in various food webs. The data come from the following publications St. Martin (Goldwasser and Roughgarden, 1993), Ythan (Hall and Raffaelli, 1991), Little Rock (Martinez, 1991).

```
links in II: 0.908123791103
links in IT: 0.0
P/R: 1.13580246914
```

2

International Trade Networks and World Trade Web

2.1 Introduction

One of the most interesting and cured datasets available in economics is the recorded set of commercial transactions between different countries. This information is available in varying detail, from the large scale of the total yearly trade between one nation and all the others, to the volume of trade of a specific (and this can be very specific) product among the various countries. Trade is another system naturally described by means of a network whenever we define the vertices as the different countries and the directed edges as the trade exchange in a specific year (with a weight given by the volume in dollars). This information is available for a rather long temporal window spanning several decades. The analysis of such a database is particularly problematic for various reasons. First, as the detail increases, the size of the dataset also increases. Second, the number of products produced and traded as well as the number of countries varies year on year. We show in the following how to deal with such problems.

One of the best sources of data is COMTRADE (http://comtrade.un.org/), that is, the United Nations Commodity Trade Statistics Database. These data are collected and organised by International Merchandise Trade Statistics (IMTS), a United Nations Division. Given the immediate applicability of the framework of complex network theory to that system, there are already many papers devoted to the analysis of the topological properties of this trade (Serrano and Boguñá, 2003; Garlaschelli and Loffredo, 2004*a*; Garlaschelli *et al.*, 2007; Tacchella *et al.*, 2012). In particular, various attempts have been made to extract economic information from such data and different approaches have been taken with respect to data (i.e. aggregated (Serrano and Boguñá, 2003) or disaggregated at various levels of refining (Hidalgo *et al.*, 2007; Hidalgo and Hausmann, 2009; Tacchella *et al.*, 2012)), and with respect to the network model used (i.e. simple (Serrano and Boguñá, 2003; Garlaschelli and Loffredo, 2004*a*), bipartite (Hidalgo and Hausmann, 2009; Tacchella *et al.*, 2012), and weighted, multinetwork (Barigozzi *et al.*, 2010)).

Actually, such a system allows us to introduce a first differentiation in the study of complex networks. Indeed, when studying trade, we have the system being composed of two distinct entities: the countries (in the order of hundreds) and the products made by them (in the order of thousands at the first four digit classifications, i.e. at a reasonably detailed specification which distinguishes between apples and bananas in the class of fresh fruits). Various applications are possible and we present them in this chapter.

Data Science and Complex Networks. First Edition. Guido Caldarelli and Alessandro Chessa.

1. The first immediate study is the analysis of relations between different countries as specified by their trade.
2. Alternatively, we can also inspect the products. This can be done by representing the system as a multinetwork (Wasserman and Faust, 1994), where vertices (countries) are connected by edges (products) of different natures (Barigozzi *et al.*, 2010).
3. Finallly, we can decide to represent the same information by means of a bipartite network (i.e. a network that can split into two sets, where the edges connect only elements of the first set with elements of the second) made of countries (whose number is N_c) and products (whose number is N_p). This can be arranged in a non-square matrix of size $N_c \times N_p$ where the entries n_{cp} are the production of product p made by the country c .

The results from these questions shed some light on basic economic quantities such as reciprocity in trade, to more complex intangible quantities such as capabilities of countries (Hidalgo and Hausmann, 2009), which are hidden behind the production and export of goods and ultimately determine the progress of wealth in a nation (at least when measured by its gross domestic product (GDP)). In particular, an indirect knowledge of these capabilities and how they can be inferred from data, is a good predictor of what kind of products that country will produce in the future.

In this chapter we shall start by giving a description of the database, and how to download and organise the data. We then continue by taking the most aggregated version of the WTW at the level of nations and then proceed using successive refinements. Codes, data and/or links for this chapter are available from **http://book.complexnetworks.net**.

2.2 Data from COMTRADE

In the following we shall use United Nations Commodity Trade Statistics Database (UN COMTRADE) data. This database is derived from National Statistical Offices and covers trade from 1962. It is possible to download a series of data where the following are specified: product, producer (reporter), buyer (partner), year. It is impossible to set "all" in all four fields as files are too large to be used in a batch window. Even if, *"Free unlimited access to UN COMTRADE is available for all users on the website of the United Nations Statistics Division (UNSD)"*[1] in practice (for technical reasons) there is a *"download limit of 50,000 data records per query"*. In any case there is no limit on the number of queries a user can make. *"UN COMTRADE offers Premium access, which allows for downloads of more than 50,000 records and for the use of advanced functions of UN COMTRADE. Premium access is payable"*.

This collection of data is used by international organizations such as the World Bank Organization (WBO), International Trade Centre (ITC), World Trade Organization (WTO), and the United Nations Conference on Trade and Development (UNCTAD). Sometimes these organisations complement the original data with different datasets or cure the database, thereby creating other sources of downloads. For example:

[1] Taken from http://comtrade.un.org/db/default.aspx.

- The international Trade Centre has import/export databases publicly available at http://www.intracen.org/ByCountry.aspx, where, by selecting a single country (e.g. Algeria) we have access to the relative web page: http://www.intracen.org/country/algeria/
- The French research center in international economics (CEPII) published the BACI International Trade Database at the product level (Gaulier and Zignago, 2010). From the site description, *"BACI is constructed using an original procedure that reconciles the declarations of the exporter and the importer"*. In particular, *"First, as import values are reported CIF (cost, insurance and freight) while exports are reported FOB (free on board), CIF costs are estimated and removed from imports values to compute FOB import values. Second, the reliability of country reporting is assessed based on the reporting distances among partners. We decompose the absolute value of the ratios of mirror flows using a (weighted) variance analysis, and an index is built for each country. These reporting qualities are used as weights in the reconciliation of each bilateral trade flow twice reported."*
- The World Trade Organization (WTO) is developing a software to access COMTRADE and other data. This software, called World Integrated Trade Solutions (WITS), allows aggregation of information from other dataset, as such, for example, tariff or other legal data, and it also allows one to realise simulations on the effect of change, such as tariff cuts etc. From the web site[2] it is possible to download data and software following a (free of charge) user registration.
- From the *Journal of Conflict Resolution* (Gleditsch, 2002) where a list of the data is available in printed form.

2.2.1 Product classification

The classification of products is a rather complicated issue. Indeed one has to define the rules of the taxonomy when aggregating them. For example, we may consider origin, form, production method, or use of any particular good as different classification criteria. To add difficulty, levels of detail on the above criteria for a particular good can be very different from country to country. Finally, technological progress constantly challenges any classification applied (e.g. smartphones are phones, music players, or mini computers?). It is therefore not surprising that various initiatives have been attempted and an international team is required in order to find a proper standard. Below we report some information taken from the web sites of the various datasets.

- From the historical background reported in United Nations publications[3] we know that the first tentative international classification was made by the League of Nations, which produced the Minimum List of Commodities for International Trade Statistics (League of Nations, 1938 (II.A.14; and corrigendum, 1939)). With the new United Nations organisation, this work was continued and expanded in the first United Nations Standard International Trade Classification (SITC) in 1950. In the first instance such classification was based on the material from which the products were made. Successive classifications rearranged products also according

[2]http://wits.worldbank.org/wits/

[3]http://unstats.un.org/unsd/trade/sitcrev4.htm

to the stage of fabrication or their industrial origin. With time different classification procedures started to become more effective: they took into consideration things like the processing stage, market practices, material used in production etc. This originated a series of revisions up to the current (4th) revision that was presented in 2006.

- HS, The Harmonized Commodity Description and Coding System has been developed and maintained by the World Customs Organization (WCO). The classification can be bought from the WCO web site and it is made from about 5000 commodity groups. In this list every good is identified by a six digit code. These codes are arranged in a hierarchical structure where the first few digits correspond to a broad series of goods and extra digits are then added to account for legal information (on customs tariffs increasing the digits from 6 to 8) and for statistical purposes (increasing from 8 to 10).

These two basic classifications have many correlations of structure. For example, by employing the headings of HS as building blocks, the United Nations Statistical Office, in consultation with experts from other governments, interested international organisations, and expert groups, produced a third and a fourth revision of SITC, while taking into account the need for continuity with the previous versions. Very good mapping between the various nomenclatures for the same product is available from the WITS web page[4].

UN COMTRADE, the most important source of trade information, has used SITC since 1962 and HS since 1988. As an example of SITC classification we list here the sections corresponding to the first digit (indicated as "sections" in the web site)

0. Food and live animals
1. Beverages and tobacco
2. Crude materials, inedible, except fuels
3. Mineral fuels, lubricants, and related materials
4. Animal and vegetable oils, fats, and waxes
5. Chemicals and related products, n.e.s.
6. Manufactured goods classified chiefly by material
7. Machinery and transport equipment
8. Miscellaneous manufactured articles
9. Commodities and transactions not classified elsewhere in the SITC

In this classification, we have section 0 with nine divisions (01= *"Meat and meat preparations"*), a total of 36 groups (every division has some: 01 has for instance four different groups), 132 subgroups (01 has 17), and finally 335 basic headings. For example "0161" is the subgroup of *"Bacon, ham and other salted, dried or smoked meat of swine"* with basic headings as 016.11 *"Hams, shoulders and cuts thereof, with bone in"*. Therefore in this classification, the lower is the number of digits used, and the broader is the category of products.

[4]http://wits.worldbank.org/wits/product_concordance.html

2.2.2 Country Classification

The situation is somewhat easier (but not too much so) for the classification of countries. Among the various sources of ambiguity that could affect time evaluation of trades we have independence of former colonies, splitting and/or unification of different countries, wars, international recognition by other countries, or simply a change of the codes adopted to describe the countries worldwide. Facing only this last issue, the standard suggested by the International Organization for Standardization is what is called the ISO 3166-1.[5] Not all databases described follow the same standard, so we list three possible situations:

- Two-letter code (ISO3166-1 alpha-2). The same for the internet domains. Mostly used for practical reasons, it is less immediate to associate a country with its code (Italy → IT, France → FR, Gibraltar → GI).
- Three-letter code (ISO3166-1 alpha-3). Three-letter country codes which allow a better visual association between the codes (Italy → ITA, France → FRA, Gibraltar → GIB).
- Three-digit numeric (ISO3166-1 numeric). The most practical for countries not using the latin alphabet. No clue on what is what (Italy → 380, France → 250, Gibraltar → 292).

Countries may also change name, and it is for this reason (as well as the foregoing) that several revisions have been made.

2.3 Projecting and symmetrising a bipartite network

The first analysis that can be done is related to the total production and export of a single country against all the others. In this system, the countries are the vertices of the graph and the total export from country i to country j is a weighted edge. This graph can be indicated by the adjacency matrices A^I and A^E for imports and exports. The first problem is that (probably because of different accounting procedures) the export data $i \to j$ of the total export from i to j does not exactly match the import data $j \leftarrow i$ of the import of j from i. In other words $A^E_{ij} \neq A^I_{ji}$. Even by restricting to an unweighted version of the network, an edge is drawn only if the import/export is relevant for the country of origin, notwithstanding the fact that the relevance could be different for the counterpart. At the coarser level of aggregation only the most important products are indicated for a particular country and in the first analysis of the WTW (Serrano and Boguñá, 2003) this problem has been solved by symmetrisation of the two datasets.

In particular if both A^I_{ij} and A^E_{ji} are different from zero,

$$\begin{cases} A^I_{ij} = \frac{1}{2}(A^I_{ij} + A^E_{ji}) \\ A^E_{ij} = \frac{1}{2}(A^E_{ij} + A^I_{ji}) \end{cases} \tag{2.1}$$

and

[5] http://www.iso.org/iso/country_codes

$$\begin{cases} A^I_{ij} = A^E_{ji} \\ A^E_{ij} = A^I_{ji} \end{cases} \tag{2.2}$$

otherwise.

Network symmetrisation

```
def net_symmetrisation(wtn_file, exclude_countries):
    DG=nx.DiGraph()

    Reporter_pos=1
    Partner_pos=3
    Flow_code_pos=2
    Value_pos=9

    dic_trade_flows={}
    hfile=open(wtn_file,'r')

    header=hfile.readline()
    lines=hfile.readlines()
    for l in lines:
        l_split=l.split(',')
        #the following is to prevent parsing lines without data
        if len(l_split)<2: continue
        reporter=int(l_split[Reporter_pos])
        partner=int(l_split[Partner_pos])
        flow_code=int(l_split[Flow_code_pos])
        value=float(l_split[Value_pos])

        if ( (reporter in exclude_countries) or \
            (partner in exclude_countries) or (reporter==partner) ):
            continue

        if flow_code==1 and value>0.0:
            #1=Import, 2=Export
            if dic_trade_flows.has_key((partner,reporter,2)):
                DG[partner][reporter]['weight']= \
                (DG[partner][reporter]['weight']+value)/2.0
            else:
                DG.add_edge(partner, reporter, weight=value)
                dic_trade_flows[(partner,reporter,1)]= \
                value #this is to mark the exixtence of the link

        elif flow_code==2 and value>0.0:
```

```
            #1=Import, 2=Export
            if dic_trade_flows.has_key((reporter,partner,1)):
                DG[reporter][partner]['weight']= \
                (DG[reporter][partner]['weight']+value)/2.0
            else:
                DG.add_edge(reporter, partner, weight=value)
                #this is to mark the exixtence of the link
                dic_trade_flows[(reporter,partner,2)]=value
        else:
            print "trade flow not present\n"

    hfile.close()

    return DG
```

In a more recent approach (Barigozzi *et al.*, 2010), it has been found that in the case of doubt it is better to use the value of the trade flow as reported by the importer.

A second problem arises when we want to consider the time evolution of this system. Especially after the Second World War and in the 1990s, new countries appeared on the scene inheriting production and export from parent countries. In order to perform a temporal analysis and to allow comparisons across different years, the solution adopted in the literature (Squartini *et al.*, 2011*a*) is to consider only a stable panel of N = 162 countries that have been present in the COMTRADE data throughout the period from 1992 to 2002. The database used at this level of aggregation is freely available (Gleditsch, 2002), with various levels of detail. In spite of the fact that various way of aggregating are possible, happily enough, the various datasets show similar properties (Fagiolo *et al.*, 2009; Garlaschelli and Loffredo, 2004*a*; Squartini *et al.*, 2011*a*).

Given these caveats, the network constructed in such a way presents a series of interesting properties.

- The structure of trade channels reveals a complex organisation similar to other networks with a skewed distribution of number of economic partners and total (in US dollars) trade.
- The degree distribution $P(k)$ of the total degree $k = k_{in} + k_{out}$ for the various countries shows a fat-tailed distribution, compatible with a power-law fit of the kind $P(k) \simeq k^{-\gamma}$, with $\gamma = 2.6$.
- The degree (number of trade partners for unweighted networks) of a country is correlated with its GDP *per capita.*
- As we shall see in the following, using the above properties we can define a model based on the GDP distribution to model the evolution of trade.
- For every vertex there is a strong correlation between the in and the out degree, even if not necessarily towards the same countries (see Section 2.4.1).
- The clustering coefficient per vertex whose degree is k, i.e. the $c(k)$ defined in (1.4) of the various countries decreases with their degree k. This behaviour is well

fitted by a function of the kind $c(k) \propto k^{\omega}$, with $\omega \propto 0.7 \pm 0.05$. The clustering coefficient averaged over the whole network is $c = 0.65$, larger than the value corresponding to a random network with similar edges and vertices (Serrano and Boguñá, 2003).
- The average degree of the neighbours of one county decreases with the degree of the country (see below assortativity) with a power-law decay $knn(k) \propto k_k^{\nu}$, with $\nu_k \simeq 0.5 \pm 0.05$ (Serrano and Boguñá, 2003).

Generate the aggregate network

```
#importing the main modules

import networkx as nx

#countries to be excluded
exclude_countries=[472,899,471,129,221,97,697,492,838,473,536,\
637,290,527,577,490,568,636,839,879,0]

#this is the magic command to have the graphic embedded
#in the notebook
%pylab inline

DG=net_symmetrisation("data/comtrade_trade_data_total_2003.csv", \
                      exclude_countries)
print "number of nodes", DG.number_of_nodes()
print "number of edges", DG.number_of_edges()

#OUTPUT
number of nodes 232
number of edges 27901
```

2.4 Neighbour quantities: reciprocity and assortativity

2.4.1 Reciprocity

The reciprocity between two vertices in a directed graph is a quantity measuring the probability of having edges in both directions between two vertices. In a complete reciprocal case if vertex A has an edge towards B, then B must reciprocate with an edge towards A. Of course we cannot define this quantity in undirected graphs. From a different point of view, we can say that this is because any undirected link establishes a reciprocal connection between the vertices involved. In the case of directed graphs the situation is different: having an edge from A to B does not ensure that also the opposite edge is present (and actually this is seldom the case in real networks). The economic meaning of such a quantity can be expressed as a measure of how much the economies of the two countries are interconnected, or rather it measures how much one depends on the other to fulfil its needs. If the graph is also weighted, reciprocity is no more the simple exchange of an edge; but there is another quantity to be reciprocated, that is, the weight of the edge. In the case considered in this chapter we can think of one edge as the export from A to B. Since the export is measured in dollars a complete reciprocity would be obtained if there is also a comparable (similar in amount) export from B to A.

Intuitively, the reciprocity in a network should take into account the likelihood with which if we have an edge from one vertex i to another vertex j we also have its counterpart going from j to i. The most obvious way to measure this probability is to check the frequency with which we have edges pointing in both directions. This is done by defining the ratio r between the number of reciprocal links $L^{\leftrightarrow}$ and the total number of links L,

$$r = \frac{L^{\leftrightarrow}}{L}. \tag{2.3}$$

When no reciprocity is present we have $r = 0$, and when every link is reciprocated we have $r = 1$. Apart from these limit cases the value of r is between 0 and 1 ($0 < r < 1$). Note that r also counts loops (self-links) as reciprocal edges; in this case, the correct normalisation would be L minus the number of loops.

As in other cases of network theory, we are typically interested not in the reciprocity itself, but in the possible deviations from an "expected" or "typical" reciprocity (the one we can measure in a directed random graph). On this point we have to bear in mind that as the density increases, also the reciprocity increases, because it becomes more probable to have reciprocal links (Garlaschelli and Loffredo, 2004*b*).

Another measure of reciprocity, ρ, can be defined based on statistical considerations. In order to do so we first start from the quantity

$$\bar{a} \equiv \frac{\sum_{i \neq j} a_{ij}}{N(N-1)} = \frac{L}{N(N-1)} \tag{2.4}$$

which measures the ratio of observed to possible directed links (link density). In this way the self-linking loops are now excluded from the normalisation L.

The new reciprocity measure ρ can now be written in the following form:

$$\rho = \frac{r - \bar{a}}{1 - \bar{a}}. \tag{2.5}$$

The new definition of reciprocity gives an absolute quantity which allows one to directly distinguish between reciprocal ($\rho > 0$) and antireciprocal ($\rho < 0$) networks, with mutual links occurring more and less often than randomly respectively.

- If all the links occur in reciprocal pairs $r = 1$ and $\rho = 1$.
- If instead $r = 0$ then $\rho = \rho_{min}$, where $\rho_{min} \equiv \frac{-\bar{a}}{1-\bar{a}}$.

This is another advantage of using ρ, because it incorporates the idea that complete antireciprocal is more statistically significant in networks with larger density, while it has to be regarded as a less pronounced effect in sparser networks.

```
#unweighted case
N=DG.number_of_nodes()
L=DG.number_of_edges()

r=float((2*L-N*(N-1)))/L

print r

#OUTPUT
0.079208630515

#weighted case
W=0
W_rep=0
for n in DG.nodes():
    for e in DG.out_edges(n,data=True):
        W+=e[2]['weight']
        if DG.has_edge(e[1],e[0]):
            W_rep+=min(DG[e[0]][e[1]]['weight'],DG[e[1]][e[0]] \
                        ['weight'])

print W,W_rep,W_rep/W

#OUTPUT
7.17766475925e+12 5.19627606057e+12 0.723950788293
```

2.4.2 Assortativity

Another two-vertices property of a network is given by its assortativity. The assortativity coefficient measures the tendency of a vertex to be connected to others with a similar/dissimilar values of degree. In the former case the network is said to be "assortative". If instead in the network (on average) the hubs are connected with vertices of low degree, the whole network is said to be"disassortative". This quantity is measured practically by computing the average degrees $K_{nn}(k)$ of the neighbours of a vertex whose degree is k. To compute this quantity let's first compute the average degree $K_{nn}(i)$ of the neighbours of a vertex i,

$$K_n n(i) = \frac{\sum_{\langle ji \rangle} k_j}{n_j}, \tag{2.6}$$

where j is a neighbour of i. Then let's average again on all the vertices i whose degree is k,

$$K_n n(k) = \frac{\sum_{i:k_i=k} K_{nn}(i)}{n_k}, \tag{2.7}$$

where n_k is the number of the vertices with degree k. In assortative networks this is an increasing function of k, while in disassortative ones it decreases with k. Another measure of assortativity is given by the assortativity coefficient that measures the correlation coefficient of the degrees of neighbour sites, normalised with the variance of the degree distribution. To compute the correlation between the value of degrees k_i and k_j of two neighbour vertices i and j we need to introduce the joined probability $P(k_i, k_j)$, related to the frequency with which we measured these two values of degrees together in a graph. At this point we can write the correlation function as

$$\langle k_i k_j \rangle - \langle k_i \rangle \langle k_j \rangle = \sum_{k_i, k_j} k_i k_j (P(k_i, k_j) - P(k_i) P(k_j)), \tag{2.8}$$

since the variance of the $P(k)$ is

$$\sigma^2 = \sum_k k^2 P(k) - (\sum_k k P(k))^2, \tag{2.9}$$

and we can write the assortative coefficient as

$$r = \frac{1}{\sigma^2} \sum_{k_1, k_2} k_1 k_2 (P(k_1, k_2) - P(k_1) P(k_2)). \tag{2.10}$$

In a network with no assortativity we can factorise the joint probability, so that

$$P(k_1, k_2) = P(k_1) P(k_2) \tag{2.11}$$

and the coefficient is zero. Positive values of r signal assortative mixing. Disassortativity corresponds to negative values of r.

Assortativity

```
#K_nn distribution
list_Knn=[]
for n in DG.nodes():
    degree=0.0
    for nn in DG.neighbors(n):
        degree=degree+DG.degree(nn)
    list_Knn.append(degree/len(DG.neighbors(n)))

#plot the histogram
hist(list_Knn,bins=12)

#basic Pearson correlation coefficient for the
r=nx.degree_assortativity_coefficient(DG)
print r
```

HS Code	Commodity	w_{ij}	Density	NS_{in}/ND_{in}	NS_{out}/ND_{out}
9	Coffee	0.309	3.3811	2.553	2.3906
10	Cereals	0.1961	5.5195	5.9919	2.5718
27	Min. Fuels	0.3057	3.3575	2.6786	3.2979
29	Org. Chem.	0.3103	3.3664	2.3579	1.6286
30	Pharmaceutical	0.3662	2.803	2.3308	1.267
39	Plastics	0.4926	2.0478	1.753	1.1385
52	Cotton	0.2864	3.5839	2.7572	2.1254
71	Prec. Metals	0.2843	3.6746	1.9479	2.6704
72	Iron	0.3081	3.3315	2.5847	1.8484
84	Nuclear Machin.	0.6195	1.6281	1.3359	1.0259
85	Electric Machin.	0.5963	1.6917	1.3518	1.0692
87	Vehicles	0.4465	2.259	1.7488	1.1105
90	Optical Instr.	0.4734	2.1492	1.5879	1.0993
93	Arms	0.1415	8.4677	6.0618	4.0279

Table 2.1 Density and node-average of topological properties of commodity-specific networks vs. aggregate trade network for the 14 most relevant commodity classes in year 2003. Ratios of the statistic value in the commodity-specific network to aggregate network are showed. Values larger (smaller) than 1.0 mean that average of commodity-specific networks is larger (smaller) than its counterpart in the aggregate network.

```
#weighted version
r=nx.degree_pearson_correlation_coefficient(DG,weight='weight', \
                                             x='out',y='out')
print r

#OUTPUT
-0.335002643638
-0.0696781960521
```

2.5 Multigraphs

Multigraphs turns out to be useful in describing trade. In practice, we can imagine having different layers of products where on every layer the vertices are connected by different kinds of edges (product). For example, the UK and the USA will then be connected by more than one edge, where every edge accounts for a different commodity traded. The aggregate weighted, directed Word Trade Web or International Trade Network is obtained by simply summing the all-commodity-specific layers. A topological analysis of the multinetwork structure allows us to assess the commodity heterogeneity in commodity-specific networks, as compared to those of the aggregate network. Following previous analysis (Barigozzi *et al.*, 2010) we can conclude that WTW (ITN) is composed of layers with rather heterogeneous properties. By considering ten of the most relevant commodities exchanged, as shown in Table 2.1, we can find a rather different behaviour with respect to degree, density, weight and strength.

Weighted networks, the strength

```
dic_product_networks={}
commodity_codes=['09','10','27','29','30','39','52','71','72','84', \
'85','87','90','93']
for c in commodity_codes:
    dic_product_networks[c]=net_symmetrisation( \
    "data/comtrade_trade_data_2003_product_"+c+".csv", \
    exclude_countries)

DG_aggregate=net_symmetrisation( \
"data/comtrade_trade_data_total_2003.csv",exclude_countries)

#rescale the weighted adjacency matrices
#aggregate
w_tot=0.0
for u,v,d  in DG_aggregate.edges(data=True):
    w_tot+=d['weight']
for u,v,d  in DG_aggregate.edges(data=True):
    d['weight']=d['weight']/w_tot
#products
for c in commodity_codes:
    l_p=[]
    w_tot=0.0
    for u,v,d  in dic_product_networks[c].edges(data=True):
        w_tot+=d['weight']
    for u,v,d  in dic_product_networks[c].edges(data=True):
        d['weight']=d['weight']/w_tot

density_aggregate=DG_aggregate.number_of_edges() / \
(DG_aggregate.number_of_nodes()*(DG_aggregate.number_of_nodes()-1.0))

w_agg=[]
NS_in=[]
NS_out=[]
for u,v,d in DG_aggregate.edges(data=True):
    w_agg.append(d['weight'])
for n in DG_aggregate.nodes():
    if DG_aggregate.in_degree(n)>0:
        NS_in.append(DG_aggregate.in_degree(n,weight='weight')/ \
                     DG_aggregate.in_degree(n))
    if DG_aggregate.out_degree(n)>0:
        NS_out.append(DG_aggregate.out_degree(n,weight='weight')/ \
```

```
                        DG_aggregate.out_degree(n))

for c in commodity_codes:
    density_commodity=dic_product_networks[c].number_of_edges() / \
    (dic_product_networks[c].number_of_nodes()* \
    (dic_product_networks[c].number_of_nodes()-1.0))
    w_c=[]
    NS_c_in=[]
    NS_c_out=[]
    for u,v,d  in dic_product_networks[c].edges(data=True):
        w_c.append(d['weight'])
    for n in dic_product_networks[c].nodes():
        if dic_product_networks[c].in_degree(n)>0:
            NS_c_in.append(dic_product_networks[c].in_degree (n, \
            weight='weight')/dic_product_networks[c].in_degree(n))
        if dic_product_networks[c].out_degree(n)>0:
            NS_c_out.append(dic_product_networks[c].out_degree(n, \
            weight='weight')/dic_product_networks[c].out_degree(n))

    print c,str(round(density_commodity/density_aggregate,4))+ \
    " & "+str(round(mean(w_c)/mean(w_agg),4))+" & "+ \
    str(round(mean(NS_c_in)/mean(NS_in),4))+" & "+ \
    str(round(mean(NS_c_out)/mean(NS_out),4))

#OUTPUT
09 0.309 & 3.3811 & 2.553 & 2.3906
10 0.1961 & 5.5195 & 5.9919 & 2.5718
27 0.3057 & 3.3575 & 2.6786 & 3.2979
29 0.3103 & 3.3664 & 2.3579 & 1.6286
30 0.3662 & 2.803 & 2.3308 & 1.267
39 0.4926 & 2.0478 & 1.753 & 1.1385
52 0.2864 & 3.5839 & 2.7572 & 2.1254
71 0.2843 & 3.6746 & 1.9479 & 2.6704
72 0.3081 & 3.3315 & 2.5847 & 1.8484
84 0.6195 & 1.6281 & 1.3359 & 1.0259
85 0.5963 & 1.6917 & 1.3518 & 1.0692
87 0.4465 & 2.259 & 1.7488 & 1.1105
90 0.4734 & 2.1492 & 1.5879 & 1.0993
93 0.1415 & 8.4677 & 6.0618 & 4.0279
```

Starting from this static analysis of every layer that can be extended to various network quantities we can now move to the analysis of cross-product correlations.

2.6 The bipartite network of products and countries

A simple way to look at the matrix M_{cp} of countries and products is to disregard for a moment information on volume of production and to transform the weighted elements M_{cp} giving the flow of US dollars in c for the trade of product p into a binary variable, specifying only whether the country is an effective producer of the product p. The criterion adopted in order to understand whether a country can be considered, or not, a producer of a particular product is the so-called Balassa's revealed comparative advantage (RCA) (Balassa, 1965). Indeed, an export relevant for a country is not necessarily relevant also for the counterpart and vice versa. Therefore, it is necessary to weigh export of a good in relation to how much of the same product is produced worldwide, i.e. $\sum_{c'} M_{c'p}$.

This must be compared with the importance of the export of single country, which is again a ratio between the total export of c (i.e. $\sum_{p'} M_{cp'}$) with respect to the global value of the exports for every country (i.e. $\sum_{c',p'} M_{c'p'}$). In formulas, we get

$$RCA_{cp} = \frac{M_{cp}}{\sum\limits_{p'} M_{cp'}} / \frac{\sum\limits_{c'} M_{c'p}}{\sum\limits_{c',p'} M_{c'p'}}. \tag{2.12}$$

We consider country c to be a competitive exporter of product p if its RCA is greater than some threshold value. In the standard economics literature this value is taken as one and small variations around such a threshold do not qualitatively change the results (Hidalgo and Hausmann, 2009).

Revealed comparative advantage

```
def RCA(c,p):
    X_cp=dic_product_networks[p].out_degree(c,weight='weight')
    X_c=DG_aggregate.out_degree(c,weight='weight')

    X_p=0.0
    for n in dic_product_networks[p].nodes():
        X_p+=dic_product_networks[p].out_degree(n,weight='weight')

    X_tot=0.0
    for n in DG_aggregate.nodes():
        X_tot+=DG_aggregate.out_degree(n,weight='weight')

    RCA_cp=(X_cp/X_c)/(X_p/X_tot)

    return RCA_cp

p='93'
```

```
c=381
print RCA(c,p)

#OUTPUT
2.10470555164
```

Once we have the data in the form of a binary matrix, we can extract features by means of spectral theory. In order to do so, we define two complementary graphs corresponding to projection of the original bipartite network on the country nodes and on the products nodes. In that way the projected graphs with respectively N_c and N_p nodes are homogeneous with respect to the two different types of nodes. The easiest way to perform this projection is to consider the following two matrix products:

$$\begin{aligned} C &= MM^T \\ P &= M^T M, \end{aligned} \tag{2.13}$$

where M^T is the transposed matrix and the square matrices C and P define the country–country network and the product–product network. The element $C_{cc'}$ defines the weight associated to the link between countries c and c' in the country–country network. Analogously $P_{pp'}$ gives the weight of the link between products p and p' in the product product network. These weights have an interesting interpretation: if we write explicitly the expression of a generic element of the C matrix according to (2.13), we have that $C_{cc'} = \sum_p M_{cp} M_{c'p}$. Therefore the element $C_{cc'}$ (since M_{cp} is a binary unweighted matrix) counts the number of products exported by both countries c and c'. In a similar way the the element $P_{pp'}$ counts the number of countries which export both products p and p'. The diagonal elements C_{cc} and P_{pp} are respectively the number of products exported by the country c and the number of exporters of the product p.

Bipartite Networks. As shown in Fig. 2.1 bipartite networks are networks composed of two distinct sets, where links connect only elements of one set with elements of the other. Typically, bipartite networks arise in social systems where one set is formed by people and the second set by the object of their collaboration. Among the many examples we have:

- the actor–movie network, where actors are the nodes in one set, and they are connected to the films in which they played, forming the nodes in the second set (Albert and Barabási, 2002);
- scientist networks that can be defined as having in one set scientists and in the other their papers (Newman, 2003);
- directors of companies and the boards in which they sit (Caldarelli *et al.*, 2004);
- and finally countries of the world and the products they produce (Caldarelli *et al.*, 2012; Tacchella *et al.*, 2012).

Any two columns of such a dataset can be adopted to define a bipartite network, the properties of which could reveal patterns and regularities worthy of attention. More generally bipartite graphs can be defined whenever it is possible to partition the graph

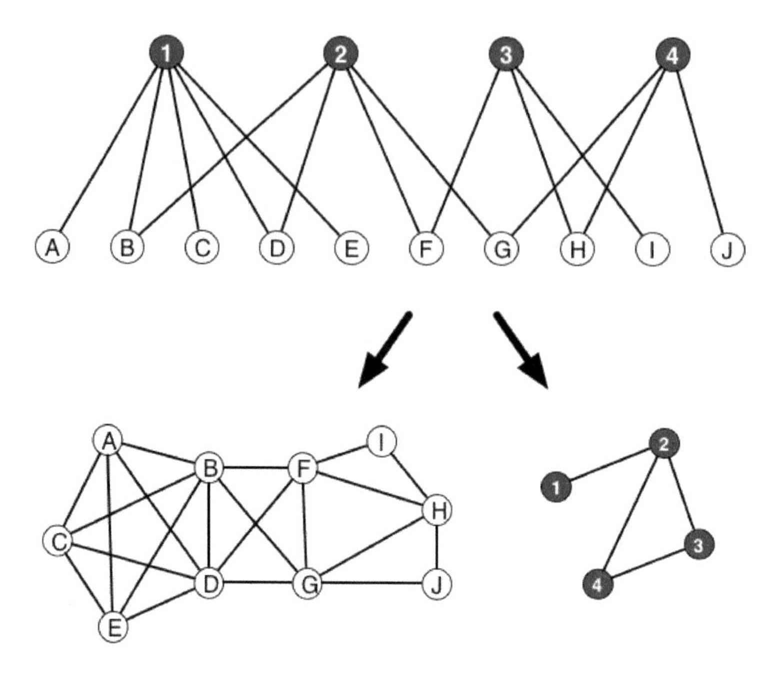

Fig. 2.1 An example of a bipartite network with the two possible projections. Note that an edge between vertex i and j in the projection can be weighted. The weight is given by the number of vertices of the other set connected to both i and j.

into two sets. This is always possible whenever the graph has no odd cycles, as shown in Fig. 2.2. The standard way to study such structures is by restricting to one of the two sets. In this way we can build a network of actors where the connection is given by having played in one or more films, or a network of films if they share one or more actors. We can keep track of the number of connections by considering a weighted graph, so that one connection between two actors is an integer number representing the total of the films they have in common. Some of the total information is then "projected" in the subspace of one of the two sets. Reasons for doing this are, for example, the need to assess the centrality (how crucial it is for the structure of the whole network) of one node with respect to the others or to determine clusters of vertices that share similar properties (community detection).

Computing the bipartite network and projections

```
import numpy as np

num_countries=DG_aggregate.number_of_nodes()
num_products=len(commodity_codes)
```

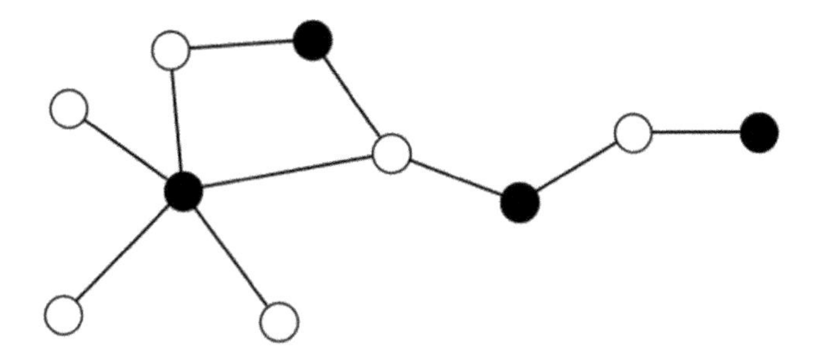

Fig. 2.2 An example of a graph that can be partitioned into two sets of white and black vertices.

```
#generate array indices
country_index={}
i=0
for c in DG_aggregate.nodes():
    country_index[c]=i
    i+=1

M=np.zeros((num_countries,num_products))

for pos_p,p in enumerate(commodity_codes):
    for c in dic_product_networks[p].nodes():
        if RCA(c,p)>1.0:
            M[country_index[c]][pos_p]=1.0
    print "\r"

C=np.dot(M,M.transpose())
P=np.dot(M.transpose(),M)

print C
print P

#OUTPUT
[[ 2.  1.  1. ...,  0.  1.  1.]
 [ 1.  4.  2. ...,  0.  3.  1.]
 [ 1.  2.  3. ...,  0.  2.  1.]
 ...,
 [ 0.  0.  0. ...,  1.  0.  0.]
 [ 1.  3.  2. ...,  0.  4.  1.]
 [ 1.  1.  1. ...,  0.  1.  2.]]
[[ 83.  27.  28.   4.   6.   6.  29.  31.  20.   1.  3.  3.  5. 12.]
```

```
[ 27.  59.  19.   4.   4.   8.  27.  18.  19.   5.  3.  7.  3. 12.]
[ 28.  19.  71.   4.   2.   7.  20.  16.  14.   3.  4.  4.  1.  9.]
[  4.   4.   4.  20.   9.   9.   2.   6.   5.   5.  4.  3.  7.  7.]
[  6.   4.   2.   9.  27.  15.   7.   6.  10.   9.  3.  8.  9. 10.]
[  6.   8.   7.   9.  15.  37.  10.   7.  15.  10. 10.  8.  9. 11.]
[ 29.  27.  20.   2.   7.  10.  69.  19.  18.   4.  5.  7.  5. 14.]
[ 31.  18.  16.   6.   6.   7.  19.  57.  10.   4.  3.  4.  6.  9.]
[ 20.  19.  14.   5.  10.  15.  18.  10.  56.   7.  7. 12.  2. 15.]
[  1.   5.   3.   5.   9.  10.   4.   4.   7.  26. 12.  9.  7.  6.]
[  3.   3.   4.   4.   3.  10.   5.   3.   7.  12. 26.  5.  8.  6.]
[  3.   7.   4.   3.   8.   8.   7.   4.  12.   9.  5. 27.  5. 11.]
[  5.   3.   1.   7.   9.   9.   5.   6.   2.   7.  8.  5. 20.  5.]
[ 12.  12.   9.   7.  10.  11.  14.   9.  15.   6.  6. 11.  5. 38.]]
```

3
The Internet Network

3.1 Introduction

One of the reasons behind the fact that network science has developed quickly over recent years is because of its timely description of new phenomena appearing in the present world. Indeed, the internet revolution has shaped an unprecedented society where the pervasive presence of computer-based services has changed completely the way in which we live our lives. Internet services such as the Web itself, but also Wikipedia, Facebook, and Twitter allow the exchange of information stored in servers and connected by a web of physical links. In spite of its use in a wider sense, the "Internet" is technically only the physical layer of PCs, computers, and servers connected by cables. Born to be the skeleton of a communication service between different parts of the USA in case of wartime attack, it soon became a way to connect universities and research institutes, to exchange scientific information, and later on it was exploited for its commercial uses.

The initial growth of the Internet was planned (Baran, 1964) as early as 1964, as a structure able to survive the destruction of one of its nodes, and the protocol of communications adopted (as for e-mails) was designed accordingly.

One of the most successful applications on the internet was the World Wide Web, which made possible the growth and development of a pletora of other services as, for example, Wikipedia, Facebook, Twitter, and all the other social networks.

All these various sectors are interacting with each other and reshaping their structure accordingly. Tim Berners Lee, coined the term GGG (giant global graph) to refer to the next revolution where all the information produced and stored in various services will be aggregated, categorised, and distributed in various formats according to the user's need (Berners Lee, 2007). Codes, data and/or interesting links for this chapter are available from **http://book.complexnetworks.net**.

3.2 Data from CAIDA

The Internet is the set of the various computers worldwide, connected by cables, servers etc, it indicates a physical framework. With the sentence "Looking for something on the Internet" we typically mean browsing a web site stored in one of those computers. The existence of this network is due to military research first (how to build a network for communication able to work after bombing and destruction of some of its parts), while scientific needs (how to efficiently share resource and information) appeared only late. Access to the system is not regulated by central administration and it has been made possible by non-proprietary standardized rules of communication known as the Internet

Protocol Suite (TCP/IP), fixed as early as 1982. This standard has been set up in such a way as to be able to operate independently from the hardware available. Another strength of the Internet is the idea of the redundancy of connections (Baran, 1960). In the spirit of the author "A communications network that uses a moderate degree of redundancy to provide high immunity from the deleterious effects of damage of relay centres. The degree of redundancy needed is shown to be determined primarily by the amount of damage expected" This idea was successfully applied in the first planning of the Internet, where a little redundancy was present. After that, however, the structure grew with no or only limited planning and, as a result, there is no complete map of the Internet available. Nevertheless, we can use probes to determine parts of the structure. "Traceroute" is a troubleshooting application that traces the path data takes from one computer to another. While its primary aim is to detect functionality of connection, it is often used to map the Internet system. This application shows the number of hops that the data makes before reaching the host (plus extra information on how long each hop takes). These "hops" count the intermediate devices (like routers) through which data must pass between source and destination, rather than flowing directly over a simple cable. Each device along the data path constitutes a hop, or in other words is a vertex in the graph. Therefore a hop count gives the distance of two nodes in the Internet network.

In practice, by using this application (available in Linux, Mac, and Windows OS) from our Internet address, we can discover the paths connecting us to any target destination. Various projects have set up repositories of traceroute data,[1] while the most comprehensive repository is based in CAIDA (Center for Applied Internet Data Analysis).

By collecting many paths from a given address we can realise a map of how the Internet is seen by a particular observer. The result, shown in Fig. 3.1, is a map of the structure realised by the internet mapping project. At the time of writing a complete map of all connections is not available, but researchers probe constantly the portions of system that are available to obtain an as accurate as possible Internet cartography. Visualisation of the Internet as well as that of any other network is particularly important. Not by chance, in many Indo-European languages, are "to see" and "to know" are obtained from the same root *wid (producing a series of words with one or other meaning, if not both, as in Latin *video/videor*, German *wissen*, Greek (ϝ-)ιδέα, English *wise*, Serbian *vid*, Sanskrit *vid*, Czech *vidět/vědět*), not to mention the ancient Greek (ϝ-)οἶδα (I know because I have seen). To "know" about networks it is then crucial to have reliable codes to visualise them in order to disentangle their complexity. This is the case with the Internet where the system is so large that only local portions are available. Internet data from traceroute produces mostly tree-like structures whose complexity can easily be simplified using appropriate visualisation tools. The best source of data for the Internet is from the Center for Applied Internet Data Analysis (CAIDA), based at the University of California's San Diego Supercomputer Center. CAIDA is "a collaboration of different organisations in the commercial, government, and research sectors investigating practical and theoretical aspects of the Internet in order to:

[1] https://labs.ripe.net/datarepository/data-sets/iplane-traceroute-dataset

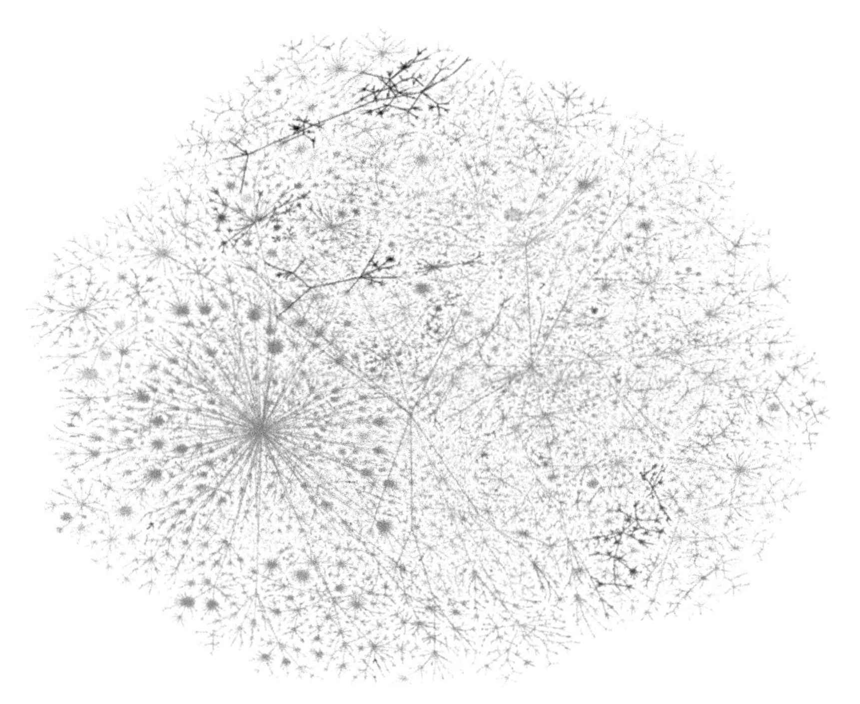

Fig. 3.1 A snaphot of the global Internet map, realised by the Internet Mapping Project (http://cheswick.com/ches/map/) on August 1998.

- provide macroscopic insights into Internet infrastructure, behavior, usage, and evolution,
- foster a collaborative environment in which data can be acquired, analyzed, and (as appropriate) shared,
- improve the integrity of the field of Internet science,
- inform science, technology, and communications public policies."

For this reason it represents a crucial place for data collection and analysis for our book

3.2.1 Visualisation

The first thing we want to show is how to visually represent a graph. Indeed, most of the important characteristics of complex networks can be spotted by just looking at them. A good visualisation algorithm, should stress the different roles of the vertices whenever it is possible to show them all. While the same graph can be drawn in a variety of forms, almost invariably the best visualisation is the one that reduces the number of crossing edges.

As in almost all cases of complex networks applications, the graphs are rather sparse, therefore, it is not particularly efficient to keep in the memory the whole adjacency matrix; rather a better choice is to consider the list of edges. For this reason, many of the available softwares for graph visualization have the list of edges as input. This is the case with the Pajek software (http://pajek.imfm.si/) which is particularly user friendly. In this case the procedure is rather simple. First just put down a row listing where you list how many vertices are in the graph. Then write the command "edges list" and then the list of edges in the format, first vertex (number) second vertex (number), so that the whole file has the form

```
Vertices 143
Edgeslist*
1  3
2  4
3  6
....
```

Other and more sophisticated desktop applications, like Gephi (http://gephi.github.io/), work with files written with a similar structure. Here we use directly the capabilities of the Python language and its graphical module Matplotlib. As a base to introduce the various centrality measures we will start from a simple and relatively small network freely available in Wikipedia.[2]

Here is a simple way to visualise this graph with Matplotlib:

Network from SVG with the best node positioning

```
import networkx as nx
from BeautifulSoup import BeautifulSoup

def Graph_from_SVG(stream):

    G=nx.Graph()

    attrs = {
        "line" :  ["x1","y1","x2","y2"]
    }

    op = open(stream,"r")
    xml = op.read()

    soup = BeautifulSoup(xml)

    count=0
    node_diz={}
    pos={}
    for attr in attrs.keys():
        tmps = soup.findAll(attr)
        for t in tmps:
```

[2] http://commons.wikimedia.org/wiki/File:Graph_betweenness.svg

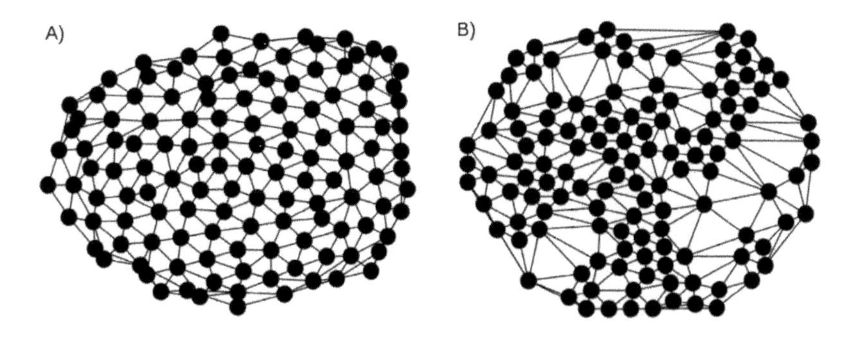

Fig. 3.2 A) The Graphviz layout for he test_graph considered in the text and B) the optimal positioning to check for node centrality measures.

```
            node1=(t['x1'],t['y1'])
            node2=(t['x2'],t['y2'])
            if not node_diz.has_key(node1):
                node_diz[node1]=str(count)
                pos[str(count)]=(float(node1[0]),float(node1[1]))
                count+=1
            if not node_diz.has_key(node2):
                node_diz[node2]=str(count)
                pos[str(count)]=(float(node2[0]),float(node2[1]))
                count+=1
            G.add_edge(node_diz[node1],node_diz[node2])
    #save the graph in an edge list format
    nx.write_edgelist(G, "./data/test_graph.dat",data=False)

    return G,pos
```

The "draw" command now presents more parameters. In particular it is possible to set the position of each node, including the "pos" parameter that is a simple Python dictionary of the type: $\{u'41' : (603.27, 84.68), \ldots\}$ where the key is the node id and the value is a couple ("tuple") formed by the two-dimensional coordinates. In the specific case we first use the graphviz layout that is able to nicely distribute nodes. Moreover the parameter "node_size" adjusts the size of the circle representing the node in order to show a more readable graph. Finally the "savefig" function stores the figure snapshot in a local file in a "png" format and with a dpi (dots per inch) equal to 200. The result is presented in Fig. 3.2A.

Even if the Graphviz layout is able to render the graph in an intelligible way, we will adopt the optimal node positions from the original SVG file with the following procedure, and at the same time we will extract from the same file its topological structure.

Visualisation tools

```
#getting the network in the SVG format
file="./data/test_graph.svg"
(G,pos)=Graph_from_SVG(file)

#plot the optimal node distribution
nx.draw(G, pos, node_size = 150, node_color='black')
#save the graph on a figure file
savefig("./data/test_network_best.png", dpi=200)

#plotting the basic network
G=nx.read_edgelist("./data/test_graph.dat")
graphviz_pos=nx.graphviz_layout(G)
nx.draw(G, graphviz_pos, node_size = 150, node_color='black')
#save the graph on a figure file
savefig("./data/test_network_graphviz.png", dpi=200)

#OUTPUT IN THE FIGURE
```

The rationale behind this algorithm is to tag the links using the coordinates as a unique id for the node names. In this way we are able both to reconstruct the topological structure of the graph from the SVG directives drawing the lines and also retrieve the exact node positions (see Fig. 3.2B for the final result).

3.3 Importance or centrality

The centrality of a vertex or edge is generally perceived as a measure of the importance of this element within the whole network. There are various ways to address these two issues and for that reason there are different measures of network centrality available. As shown in Fig. 3.3, different centrality measurements determine different sets of vertices(Perra and Fortunato, 2008; Boldi and Vigna, 2014).

3.3.1 Degree centrality

One "local" measure of centrality is to look for the vertices with the largest degrees. Indeed, being very well connected they are probably often visited by anyone travelling on the graph. This quantity called "degree centrality" is local since it can only be computed by checking the vertex itself and, in most cases, it represents a fast and reasonably accurate quantity to describe the importance of vertices in a graph. We can very quickly get the degree values for all the nodes through the following NetworkX function:

Degree sequence

```
degree_centrality=nx.degree(G)
print degree_centrality

#OUTPUT
{u'24': 7, u'25': 6, u'26': 4, u'27': 7, u'20': 7, u'21': 6,
 u'22': 4, u'23': 4, u'28': 6, u'29': 6, u'0': 5, u'4': 5,
...}
```

We then generate, plot, and save the figure (see Fig. 3.3A).

```
l=[]
res=degree_centrality
for n in G.nodes():
    if not res.has_key(n):
        res[n]=0.0
    l.append(res[n])

nx.draw_networkx_edges(G, pos)
for n in G.nodes():
    list_nodes=[n]
    color = str( (res[n]-min(l))/float((max(l)-min(l))) )
    nx.draw_networkx_nodes(G, {n:pos[n]}, [n], node_size = 100, \
    node_color =
    color)

savefig("./data/degree_200.png",dpi=200)
```

In the visualisation process we introduce new functions and paramenters. In the first place we draw just the edges with the function "draw_networkx_edges", following the precise positions through the variable "pos". Then we plot the nodes (one by one) generating a colour code proportional to the centrality, as a float between 0 and 1, normalised according to the difference between the maximum and the minimum degree centrality attained in the test graph. We will follow a similar procedure for all of the other centrality measures.

3.3.2 Closeness centrality

A non-local definition of centrality is based on the notion of distance. The quantity is non-local since we need to inspect the whole graph to compute it. The lower the distance from the other vertices the larger is the closeness. In this way we get for vertex i, the closeness c_i formula

$$c_i = \frac{1}{\sum_{j \neq i} d_{ij}}. \tag{3.1}$$

Of course the above formula makes sense only for sites i, j in the connected component (otherwise we assume that d_{ij} is infinite). For networks that are not strongly connected, a viable alternative is harmonic centrality:

$$c_i^h = \sum_{j \neq i} \frac{1}{d_{ij}} = \sum_{d_{ij} < \infty, j \neq i} \frac{1}{d_{ij}} \tag{3.2}$$

which replaces the implicit arithmetic mean of closeness with a harmonic mean (Boldi and Vigna, 2014). To compute these centrality measures we need a function that computes all the distances from a root node. Here we can use the BFS algorithm that we introduced in Chapter 1. This function will return the list of distances as a list of "tuples": (edge, distance):

Distance function

```
def node_distance(G,root_node):
    queue=[]
    list_distances=[]
    queue.append(root_node)
    #deleting the old keys
    if G.node[root_node].has_key('distance'):
        for n in G.nodes():
            del G.node[n]['distance']
    G.node[root_node]["distance"]=0
    while len(queue):
        working_node=queue.pop(0)
        for n in G.neighbors(working_node):
            if len(G.node[n])==0:
                G.node[n]["distance"]=G.node[working_node] \
                ["distance"]+1
                queue.append(n)
    for n in G.nodes():
        list_distances.append(((root_node,n),G.node[n]["distance"]))
    return list_distances
```

Then the closeness computes this function for all the nodes in our test network and plots it (see Fig. 3.3B).

Closeness

```
norm=0.0
diz_c={}
```

```
l_values=[]
for n in G.nodes():
    l=node_distance(G,n)
    ave_length=0
    for path in l:
        ave_length+=float(path[1])/(G.number_of_nodes()-1-0)
    norm+=1/ave_length
    diz_c[n]=1/ave_length
    l_values.append(diz_c[n])

#visualization
nx.draw_networkx_edges(G, pos)
for n in G.nodes():
    list_nodes=[n]
    color = str((diz_c[n]-min(l_values))/(max(l_values)- \
                                          min(l_values)))
    nx.draw_networkx_nodes(G, {n:pos[n]}, [n], node_size = \
                           100, node_color = color)

savefig("./data/closeness_200.png",dpi=200)
```

3.3.3 Betweenness centrality

Another "non-local" way to measure the importance of one vertex or edge is to check how often we visit it when walking on the network. When we consider all the distances among vertices in the network, we cross some "in between" sites more than once. The more often we pass through a certain site, the larger is its "betweenness" :

$$b(i) = \sum_{\substack{j,l=1,n \\ i \neq j \neq l}} \frac{\mathcal{D}_{jl}(i)}{\mathcal{D}_{jl}}, \tag{3.3}$$

where $\mathcal{D}_{jl}$ is the total number of different shortest paths (distances) going from j to l and $\mathcal{D}_{jl}(i)$ is the subset of those distances passing through i. The sum runs over all pairs with $i \neq j \neq l$. The larger the degree of a vertex, the larger is on average its betweenness; the two quantities are correlated and it is possible to connect the properties of the betweenness distribution to that of the degree distribution (Goh, Kahng and Kim, 2001; Barthélemy, 2004). See Fig. 3.3C for the result.

Betweenness

```
list_of_nodes=G.nodes()
num_of_nodes=G.number_of_nodes()
```

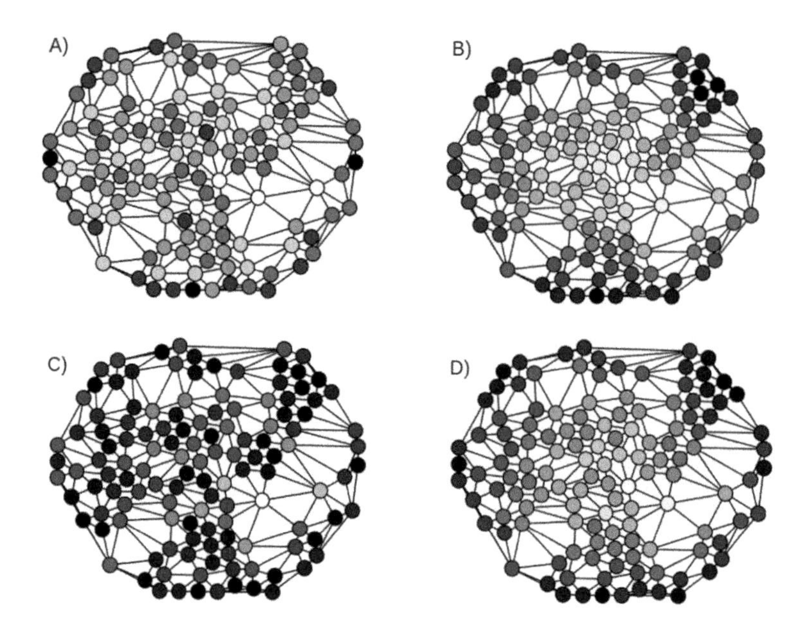

Fig. 3.3 Examples of A) degree centrality, B) closeness centrality, C) betweenness centrality, D) eigenvector centrality of the same graph.

```
bc={} #we will need this dictionary later on
for i in range(num_of_nodes-1):
    for j in range(i+1,num_of_nodes):
        paths=nx.all_shortest_paths(G,source=list_of_nodes[i], \
                                    target=list_of_nodes[j])
        count=0.0
        path_diz={}
        for p in paths:
            #print p
            count+=1.0
            for n in p[1:-1]:
                if not path_diz.has_key(n):
                    path_diz[n]=0.0
                path_diz[n]+=1.0
        for n in path_diz.keys():
            path_diz[n]=path_diz[n]/count
            if not bc.has_key(n):
                bc[n]=0.0
            bc[n]+=path_diz[n]

#visualization
```

```
l=[]
res=bc
for n in G.nodes():
    if not res.has_key(n):
        res[n]=0.0
    l.append(res[n])

nx.draw_networkx_edges(G, pos)
for n in G.nodes():
    list_nodes=[n]
    color = str( (res[n]-min(l))/(max(l)-min(l)) )
    nx.draw_networkx_nodes(G, {n:pos[n]}, [n], node_size = 100, \
                           node_color = color)

savefig("./data/betweenness_200.png",dpi=200)
```

Betweenness centrality is particularly useful in the case of community detection. Indeed, it is a measure of the "bridging" properties of one vertex/edge so that edges with large betweenness are likely to bridge different communities. Following this idea, if we remove them we can isolate the communities present in the graph (Girvan and Newman, 2002). The idea is to recursively compute the betweenness of the various edges in the network and to remove those with the largest values. In this way, isolated communities emerge from the web of connections. By iterating this procedure, the edges are removed one by one and the vertices become disconnected.

In our example algorithm, since the graph is small, we use a trivial procedure that calculates all possible minimal paths between nodes, but more refined algorithms have been introduced (Brandes, 2001) for larger graphs.

3.3.4 Eigenvector centrality

Finally we introduce a spectral centrality measure. It is based on the spectral properties of the adjacency matrix A (other measures are based on simple functions of it). The starting point is to define the centrality of a vertex i as the average of the centrality of its neighbours, i.e.

$$c_i = \frac{1}{\lambda} \sum_{j=1,N} a_{ij} c_j. \tag{3.4}$$

In its vectorial form the above equation can be written as

$$A\vec{c} = \lambda\vec{c}. \tag{3.5}$$

That is, the centrality is an eigenvector of the adjacency matrix A, where λ is the corresponding eigenvalue. To have a physical sense the above eigenvalue must be real, but in general this is not always ensured. To partly overcome these problems it is a good choice to take λ as the largest (in absolute value) eigenvalue of matrix A. As we see later (see Section 4.3), by using the Perron–Frobenius theorem, this means that if A is irreducible, or equivalently if the graph is (strongly) connected, then the eigenvector $\vec{c}$ is both unique and positive.

To solve the above problem numerically we use a power iteration method also known as the Von Mises iteration method. The idea is to start with a good approximation of the eigenvector related to the largest eigenvalue (dominant eigenvector), or directly from a random one, and iterate the vector coefficients according to the relation

$$b_{k+1} = \frac{Ab_k}{\|Ab_k\|}. \tag{3.6}$$

In this way, at every iteration, the vector b_k is multiplied by the matrix A and normalised. In order for a subsequence of (b_k) to converge, it is sufficient that the matrix A has an eigenvalue that is strictly greater in magnitude than its other eigenvalues and also the starting vector b_0 must have a nonzero component in the direction of an eigenvector associated with the dominant eigenvalue.

Eigenvector centrality

```
#networkx eigenvector centrality
centrality=nx.eigenvector_centrality_numpy(G)

#visualization
l=[]
res=centrality
for n in G.nodes():
    if not res.has_key(n):
        res[n]=0.0
    l.append(res[n])

nx.draw_networkx_edges(G, pos)
for n in G.nodes():
    list_nodes=[n]
    color = str( (res[n]-min(l))/(max(l)-min(l)) )
    nx.draw_networkx_nodes(G, {n:pos[n]}, [n], node_size = 100, \
    node_color = color)

savefig("eigenvetor_200.png",dpi=200)
```

The result of this procedure is illustrated in Fig. 3.3D.

3.4 Robustness and resilience, giant component

Robustness and resilience are concepts often invoked in the field of critical infrastructure, as for example with the Internet, water pipelines, and electricity grid. It is

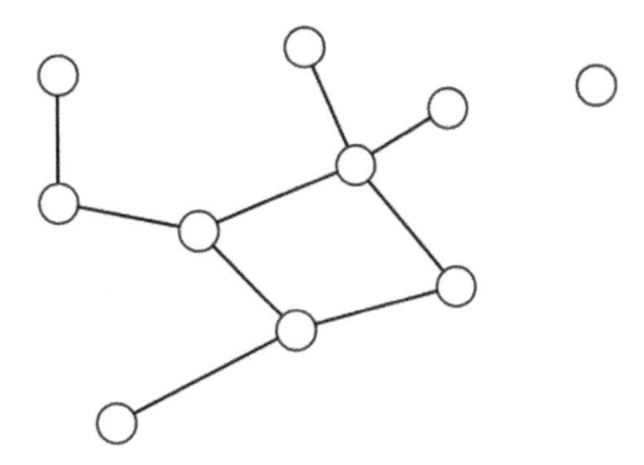

Fig. 3.4 An example of a network with two components.

important to note that they refer to distinct even if similar properties. The first quantity i.e. *robustness* iss more a static property referring to how well a system can resist an attack or failures, before being disrupted. The second quantity, i.e. *resilience* is more dynamic and describes how a system can reshape itself to avoid being disrupted.

In graph theory and network analysis, sometimes the two terms are confused and in general the "robustness" of a graph indicates the probability of remaining connected under successive removal of edges or vertices. Traditionally one of the simplest formulas for measuring robustness checks for the size of the giant connected component(Callaway, Newman, Strogatz and Watts, 2000) (i.e. the largest subset of the system which may correspond to the whole system at the beginning). When the size reduces, the network starts deteriorating. Here we present a "built-in" algorithm to compute this quantity. Networkx already has an efficient function to perform this operation, but as an example of the use of the BFS introduced previously (see Section 1.5), we code this procedure from scratch. In the first place we need a small test network with a single disconnected component to verify proper discovery of the network components (see Fig. 3.4):

Generating the graph with two components

```
G_test=nx.Graph()
G_test.add_edges_from([('A','B'),('A','C'),('C','D'),('C','E'),
                       ('D','F'), ('D','H'),('D','G'),('E','G'),
                       ('E','I')])
#disconnetted node
G_test.add_node('X')
nx.draw(G_test)

savefig("components_200.png",dpi=200)
```

Then the idea is to apply the BFS starting from a node until it discovers the component attached to it, and then progressively extract all possible components. In our case the output will be the GCC and the number of components present in the network:

Giant component through a breadth first search

```
def giant_component_size(G_input):

    G=G_input.copy()

    components=[]

    node_list=G.nodes()

    while len(node_list)!=0:
        root_node=node_list[0]
        component_list=[]
        component_list.append(root_node)
        queue=[]
        queue.append(root_node)
        G.node[root_node]["visited"]=True
        while len(queue):
            working_node=queue.pop(0)
            for n in G.neighbors(working_node):
                #check if any node attribute exists
                if len(G.node[n])==0:
                    G.node[n]["visited"]=True
                    queue.append(n)
                    component_list.append(n)
        components.append((len(component_list),component_list))
        #remove the nodes of the component just discovered
        for i in component_list: node_list.remove(i)
    components.sort(reverse=True)

    GiantComponent=components[0][1]
    SizeGiantComponent=components[0][0]

    return GiantComponent,len(components)

(GCC, num_components)=giant_component_size(G_test)
print "Giant Connected Component:",GCC
print "Number of components:",num_components

#OUTPUT
Giant Connected Component: ['A', 'C', 'B', 'E', 'D', 'I',
            'G', 'H', 'F']
```

Fig. 3.5 The GCC broken into two components after the node deleting procedure.

```
Number of components: 2
```

Now we will try to break up the GCC randomly deleting some of its nodes. To check the result visually we will apply this new function to the network defined at the beginning of this chapter (Fig. 3.5):

Breaking the GCC

```
import copy

def breaking_graph(H,node_list):
    #define the new graph as the subgraph induced by the GCC
    n_l=copy.deepcopy(node_list)
    #iterate deleting from the GCC while the graph comprises
    #one component (num_components=1)
    num_components=1
    count=0
    while num_components==1:
        count+=1
        #select at random an element in the node list
        #node_to_delete=random.choice(H.nodes())
        #select a node according to the betweenness ranking
        #(the last in the list)
```

```
        node_to_delete=n_l.pop()
        H.remove_node(node_to_delete)
        #(GCC,num_components)=giant_component_size(H)
        num_components=nx.number_connected_components(H)
    return count

(GCC, num_components)=giant_component_size(G_test)

G_GCC = G_test.subgraph(GCC)

random_list=copy.deepcopy(G_GCC.nodes())
random.shuffle(random_list)

c=breaking_graph(G_GCC,random_list)

print "num of iterations:", c

graphviz_pos=nx.graphviz_layout(G_GCC)

nx.draw(G_GCC, graphviz_pos, node_size = 200, with_labels=True)

savefig("./data/broken_component_200.png",dpi=200)

#OUTPUT
num of iterations: 1
```

Using the previous code and iterating over 1000 possible realisations of the random deleting procedure, we can get the typical number of iterations to break up the network. We will test this robustness procedure on data describing a map of the Internet at the Autonomous System (AS) level (see http://www.cosinproject.eu/extra/data/internet/nlanr.html):

Breaking up the giant connected component randomly

```
#loading the Autonomous System (AS) graph
G_AS=nx.read_edgelist("./data/AS-19971108.dat")
print "number of nodes:",G_AS.number_of_nodes(), \
"number of edges:",G_AS.number_of_edges()

(GCC, num_components)=giant_component_size(G_AS)

n_iter=1000
```

```
count=0.0
for i in range(n_iter):
    G_GCC = G_AS.subgraph(GCC)
    random_list=copy.deepcopy(G_GCC.nodes())
    random.shuffle(random_list)
    c=breaking_graph(G_GCC,random_list)
    count+=c

print "average iterations to break GCC:",count/n_iter

#OUTPUT
number of nodes: 3015 number of edges: 5156
average iterations to break GCC: 8.35
```

In this case, on average, the procedure takes approximatively eight steps to break the giant component. The robustness of the network depends on both the node deleting procedure and on the nature of the network itself. We can take out nodes randomly or follow a particular order based on some centrality measure. Moreover there are networks that are intrinsically fragile because of their topological structure (mostly regardless of the procedure adopted, e.g. trees). Previously we probed our network robustness against a random node deletion. Now we try to break it following an order that is related to a particular ranking based on the betweenness centrality measure. The general result is that following the centrality measure the network breaks down much more quickly. Here we use the betweenness centrality measure from the Networkx library. The final result is that the AS network breaks immediately, just after one node deletion.

Breaking up the giant connected component with betweenness centrality

```
import operator

G_GCC = G_AS.subgraph(GCC)

node_centrality=nx.betweenness_centrality(G_GCC, k=None, \
normalized=True, weight=None, endpoints=False, seed=None)
#node_centrality=nx.degree_centrality(G)

sorted_bc = sorted(node_centrality.items(), \
key=operator.itemgetter(1))

#selecting the ranked node list
```

```
node_ranking=[]
for e in sorted_bc:
    node_ranking.append(e[0])

c=breaking_graph(G_GCC,node_ranking)

print "num of iterations:", c

#OUTPUT
num of iterations: 1
```

4

World Wide Web, Wikipedia, and Social Networks

4.1 Introduction

Among the various services working on the Internet, one of the most successful has been the World Wide Web (WWW). In spite of the difference between the two services, even now it is common to find people interchangeably using internet (the physical device) for the WWW (the set of media and documents connected by hyperlinks). The success of the WWW has been immense. It represents the largest construction made by man, it encompasses most of world knowledge, it allows the exchange of information and feelings between people (even if physically very far apart) and it has become the environment in which different projects such as Wikipedia (an on-line encyclopaedia in almost every language of the world); Facebook (initially a system to stay in touch with a "restricted" group of friend, but now also a platform for messaging, mailing, e-commerce, and a source of news and information), and other social networks (RenRen operating in China, VK operating in Russia), and also a site for microblogging such as Twitter.

The WWW bases its success on the potential offered by the hypertext markup language (html). Thanks to this language it is possible to link documents and media with each other creating a network of information. Based on this idea, Tim Berners Lee (a collaborator of the physics laboratory CERN) decided in 1989 to put together some material in a series of pages that could be "browsed" with a specific piece of software. In such a delocalised structure it then becomes necessary to specify the location where the information is stored. This is possible thanks to the protocols of address for servers on the Internet. In any case the Internet numerical structure (i.e. an address of the kind 193.67.163.111) is rather uninformative and too restricted to keep track of all documents. As a result, WWW developed a series of addresses of the form "www.oup.com". Thanks to a hierarchical classification, an almost infinite series of documents can be mapped below this address. The mapping is stored in specific servers named "domain name servers (DNS)".

From the very beginning, many institutions, companies and private individuals put their information on line, and at the same time people worldwide started building their personal and business home pages. As the number of pages started to grow it was necessary to have a "telephone list" of the information present on the system. The first and most obvious solution was to compile a topical list. Web searching was mostly done through a page of a company named Yahoo! (yet another hierarchical officious oracle !), which was an "officious" list of links organised hierarchically. This meant that any new page had to be found and put manually into this artificial taxonomy to be present in the list. This task became more and more difficult as the numbers exploded (where and how to find all new pages?) and the content became more and more complex (how to assess the category of a web page?) to classify. Codes, data and/or links for this chapter are available from **http://book.complexnetworks.net**.

4.2 Data from various sources

4.2.1 WWW

The WWW is a classic example of big data. Often described as the largest coherent structure created by humans; actually its size (in the order of tens of billions) only refers to the static pages. Indeed, some web pages are created on demand (think of web pages for the days in a calendar) when users look for them, so that the real size of the WWW is virtually infinite. It is therefore of the utmost importance to be able to handle these series of data, and whenever possible to consider properly defined subsets of them. To that purpose we would suggest starting an exploration of the web from a set of databases collected by the Laboratory for web algorithmics of the University of Milan, Italy.

- At http://law.di.unimi.it we can find information on this site;
- http://law.di.unimi.it/datasets.php contains a series of data collected and stored in compressed form;
- http://webgraph.di.unimi.it/ contains information about the Webgraph compressed graph format and instructions on how to extract it.

On this site one can download crawls of the web of different sizes: "small" ones to test software (about 10^5 sites) to larger ones (about $10^6 - 10^7$ sites). The procedure for getting networks from these compressed files is not simple and for the benefit of the reader we have performed this task, generating a "ready to use" edge list, particularly related to a portion of the European domain name ".eu" (http://law.di.unimi.it/webdata/eu-2005/).

Code for loading the ".eu" portion of the WWW in 2005

```
import networkx as nx

#defining the eu directed graph
eu_DG=nx.DiGraph()
#retrieve just the portion of the first 1M edges of the .eu domain
#crawled in 2005
eu_DG=nx.read_edgelist('./data/eu-2005_1M.arcs', \
                       create_using=nx.DiGraph())
#generate the dictionary of node_is -> urls
file_urls=open('./data/eu-2005.urls')
count=0
dic_nodid_urls={}
```

```
while True:
    next_line=file_urls.readline()
    if not next_line:
        break
    next_line[:-1]
    dic_nodid_urls[str(count)]=next_line[:-1]
    count=count+1
file_urls.close()

#generate the strongly connected component
scc=[(len(c),c) for c in sorted( nx.strongly_connected_components \
                                (eu_DG), key=len, reverse=True)][0][1]
eu_DG_SCC = eu_DG.subgraph(scc)
```

4.2.2 Twitter

Twitter (twitter.com) is a microblogging platform, which is a service that allows its users to exchange short comments ("tweets"). Its rapid success has now made it possible to track down messages and their forwarding ("retweets") to millions of bloggers. In Twitter, each user has an account from which it is possible to write up to 140 characters in "tweets" to followers. Some users have tens of thousands of followers, others much fewer. Such "following" relationship is not reciprocal (i.e. if A follows B, not necessarily does B follow A). Twitter and Facebook are two clear cases where networks help in measuring social relationships. In particular they are a typical case of study of the new computational social science(Lazer *et al.*, 2009; Gonçalves *et al.*, 2011; Del Vicario *et al.*, 2016) Twitter provides application programming interfaces (APIs) to access tweets and information about tweets and users (https://dev.twitter.com/docs). The Python module for interacting with the Twitter API is Twython and can be reached from this link: https://twython.readthedocs.org/.

Code for the opening of tweets with the API

```
#To get your own KEYS and TOKENS visit the following page:
#https://dev.twitter.com/docs/auth/tokens-devtwittercom
#(you have to sign in before with your Twitter account)

from twython import Twython

APP_KEY='XXXXXXXXXXXXXXXXXXXXXXXXXXXXXXXXXXXXXX'
APP_SECRET='XXXXXXXXXXXXXXXXXXXXXXXXXXXXXXXXXXXXXX'
OAUTH_TOKEN='XXXXXXXXXXXXXXXXXXXXXXXXXXXXXXXXXXXXXX'
OAUTH_TOKEN_SECRET='XXXXXXXXXXXXXXXXXXXXXXXXXXXXXXXXXXXXXX'
```

```
twitter_connection=Twython(APP_KEY, APP_SECRET, \
                           OAUTH_TOKEN,OAUTH_TOKEN_SECRET)
```

From here it is now possible to get data from Twitter such as for example, the tweets from the timeline of the user.

How to get the timeline

```
#the following tweets and query results
#depend on the KEYS and TOKENS of the user

res=twitter_connection.get_home_timeline()
for t in res[:5]:
    #print the text of the first 5 tweets of the actual timeline
    print 'Text of the tweet:',t[u'text']
    #for each tweet print the mentioned users
    print 'mentions:',
    for m in t[u'entities'][u'user_mentions']:
        print m[u'screen_name'],
    print '\r'

#OUTPUT
Tweet: Vincere e vinceremo, l'ultimo scandalo del doping di
stato, non il solo eh. https://t.co/8Mzz2RmC3H
mentions:
Tweet: Ad IO SPAZIO sabato 14 novembre sar\`a ospite un amico
ed un grande professionista. @michelecucuzza
#gramigna https://t.co/wh5zdph98p
mentions: michelecucuzza
Tweet: RT @Jodinette: #LaPhraseQuiMenerve t'es contre l'euthanasie?
T'es contre le droit de mourir dans la dignit\'e!
https://t.co/0xywnSVNGU
mentions: Jodinette
Tweet: 10 days of round the clock curfew. Very harsh policy,
punishes all the people stuck there.  https://t.co/eYw1pPET02
mentions:
Tweet: Terrorismo internazionale, Merano crocevia degli aspiranti
jihadisti https://t.co/PvUP9x0chn
mentions:
```

In the following case we check information about the President of the United States, Mr. Barak Obama; in particular his location and number of followers.

How to get user information

```
res=twitter_connection.show_user(screen_name='@BarackObama')
print res
print 'location: ',res[u'location']
print 'number of followers: ',res['followers_count']

#OUTPUT
{u'follow_request_sent': False, u'has_extended_profile':
True, u'profile_use_background_image': True, u'profile_text_color':
 u'333333',
u'default_profile_image': False, u'id': 813286,
u'profile_background_image_url_https':
u'https://pbs.twimg.com/profile_background_images/
451819093436268544/kLbRvwBg.png',u'verified': True,
u'profile_location': None, u'profile_image_url_https':
u'https://pbs.twimg.com/profile_images/451007105391022080/
iu1f7brY_normal.png', u'profile_sidebar_fill_color': u'C2E0F6',
u'entities':{u'url': {u'urls':[{u'url': u'http://t.co/O5Woad92z1',
u'indices': [0, 22], u'expanded_url':
u'http://www.barackobama.com', u'display_url': u'barackobama.com'}]},
u'description': {u'urls': []}},...
...

location:  Washington, DC
number of followers:  65822369
```

After the Twitter timeline and user information one could be interested in getting a bunch of tweets related to a particular topic or hashtag. To this end the Twitter API offers a search function. In the following example we will use it to extract some tweets in which the hashtag "#ebola" appears.

Retrieving tweets with the "search" function

```
res=twitter_connection.search(q='#ebola', count=2)
for t in res['statuses']:
    print "Text of the tweet:",t[u'text']

#OUTPUT
Tweet: Read my interview on @CNN about how #SierraLeone can turn
its economy around after #Ebola https://t.co/hmeCKT5RVC
Tweet: Brazil tests man for #Ebola, puts others under observation
https://t.co/OVGSTPiqNK https://t.co/F17Xjzos09
```

Following a similar approach one could start to monitor the activity of a series of politicians in a given state and, more interestingly, the activity of all users related to these persons. Various studies have shown clearly how this information could be extremely valuable for providing an idea of the political situation of a country, especially around elections time (Eom *et al.*, 2015; Caldarelli *et al.*, 2014; Tumasjan *et al.*, 2010; DiGrazia *et al.*, 2013; Albrecht *et al.*, 2007; Adamic and Glance, 2005). An archive of some cured data for Italian political elections is available at http://www.linkalab.it/data

4.2.3 Wikipedia

Wikipedia is another service based on the WWW. It consists of a series of web pages written by a very large community of editors, on a variety of different arguments. In time it became an open access/open edit on-line encyclopaedia, whose reliability is ensured by the constant control of editors and users. The various pages are interconnected (links between existing pages are constantly being created by readers and editors) forming one of the largest thematic subnetworks of the WWW. For this reason, it has for quite a long time attracted the interest of scientists (Martin, 2011; Capocci *et al.*, 2006; Zlatić *et al.*, 2006). Interest in this subset of WWW pages is based on a series of reasons.

- Wikipedia is a well defined subgraph of the WWW; indeed it forms a thematic subset, thereby creating a natural laboratory for WWW studies.
- Over time Wikipedia has developed in different languages, so that various subsets of Wikipedia of different sizes are now available. Furthermore, Wikipedia networks allow us to test whether different cultures tend to organise web pages differently.
- All information on the Wikipedia graph is available, even its growth history, with a time stamp for any additions to the system.
- Wikipedia pages tend (where possible) to cite other Wikipedia pages, so that the whole system is contained.

In such a (extremely well connected) network it is interesting to see if the links connecting two pages (lemmas of the encyclopaedia) determine communities of concepts and ultimately define a bottom-up taxonomy of reciprocal concepts (as one would expect). For the purpose of this book it is important to note that all the data about the present shape of the network (and its growth) is publicly available and can be downloaded from

- http://dumps.wikimedia.org/

A general introduction to the subject and details of how to manage Wikipedia database files is presented on the site[1]. Dumps, varying from the largest dataset of the English version to smaller samples. We can start from the latter and then move to larger and larger datasets. In the following, for example, we shall use a small portion of Wikipedia, that "in Limba Sarda" (Sardinian), which is at the moment formed from about 4500 articles[2]. Even though the procedure of querying the Mysql database is

[1] https://meta.wikimedia.org/wiki/Data_dumps
[2] https://sc.wikipedia.org/

beyond the scope of the present book, we will sketch in the following core the main steps to extracting the hyperlink information from the dump. But in the end we will store the link structure and the page titles in a local file that we will use later on to load and to populate the proper Networkx and dictionary structures. It is possible to find the structure of the Pagelinks and Page table dumps in the following links: https://www.mediawiki.org/wiki/Manual:Pagelinks_table

https://www.mediawiki.org/wiki/Manual:Page_table

Opening the Wikipedia Sardinian dump

```
#You can skip the following cell if you don't have mysql installed
#and use directly the filesscwiki_edgelist.dat and
#scwiki_page_titles.dat you will find in the 'data' directory


#open the DB connection
#the scwiki mysql dumps scwiki-20151102-pagelinks.sql and
#scwiki-20151102-page.sql (both in the 'data' dir) have to be loaded
#in the tables "pagelinks" and "page" of the DB "scwiki_db" (to be
#created) before to launch this procedure through these commands:
#mysql -u<user> -p<password> scwiki_db< scwiki-20151102-pagelinks.sql
#mysql -u<user> -p<password> scwiki_db< scwiki-20151102-page.sql

import _mysql

scwiki_db=_mysql.connect(host="localhost",user="root", \
                         passwd="mumonkan",db="scwiki_db")

#extract the hyperlinks information with a SQL query
#from the mysql DB and storing them in a local file
scwiki_db.query("""SELECT pagelinks.pl_from, page.page_id
FROM page,pagelinks
WHERE page.page_title=pagelinks.pl_title""")
r=scwiki_db.use_result()
f=open("./data/scwiki_edgelist.dat",'w')
res=r.fetch_row()
while res!=():
    f.write(res[0][0]+" "+res[0][1]+"\n")
    res=r.fetch_row()
f.close()

#extract the title information with a SQL query
#from the mysql DB and storing them in a local file
scwiki_db.query("SELECT page.page_id,page.page_title FROM page")
```

```
r=scwiki_db.use_result()
f=open("./data/scwiki_page_titles.dat",'w')
res=r.fetch_row()
while res!=():
    f.write(res[0][0]+" "+res[0][1]+"\n")
    res=r.fetch_row()
f.close()
```

4.2.4 Wikipedia taxonomy

Since Wikipedia is a means of organising knowledge (Gonzaga *et al.*, 2001), it is interesting to check whether the structures arising from different languages and then different cultures have some sort of universality (Muchnik *et al.*, 2007; Capocci *et al.*, 2008). Furthermore the network formed by articles and hyperlinks together could provide a self-organized way to gather Wikipedia articles into categories; a classification that it is currently created upon the agreement of the whole Wikipedia community. The simplest way to create a taxonomy is by use of a tree in the shape of the Linnean taxonomy of living organisms (Linnaeus, 1735). This topic has been thoroughly studied over past years. Historically, the complexity (i.e. the fat-tailed distribution of the number of offspring at the various levels) of the structure of natural taxonomic trees from plants and animals (Willis and Yule, 1922) led to the Yule model for the growth of trees (Yule, 1925), where mutations in a population of individuals may eventually form a series of different species in the same genus.

Such a clean structure does not, unfortunately, fully apply to Wikipedia. Indeed, articles and categories will not strictly form a perfect tree, since an article or a category may happen to be the offspring of more than one parent category. For this reason the taxonomy of articles is represented in this case as a direct acyclic graph. This means that the taxonomy must be considered only as a soft partition, where the intersection between classes is different from zero. In this case one deals with (so-called) fuzzy partitions.

4.3 Bringing order to the WWW

In this section we present a short overview of the various methods that have been presented and made public to infer the importance (centrality) of pages in the WWW. Nowadays, modern search engines have (very likely) far more complicated algorithms and methods, nevertheless the original methods are still important for other cases of study and they make an excellent stage for presenting important concepts of graph theory. In the cases of studies that we present here, we define the importance of a page only topologically i.e. without entering into semantic analysis of the content of a single page. The first algorithm using such an approach was introduced in 1999 (Kleinberg, 1999) under the name HITS (Hyperlink-Induced Topic Search).

4.3.1 HITS algorithm

As a first approximation, let's make a basic differentiation of pages into two categories:

- *authorities* i.e. pages that contain relevant information (train timetable, food recipes, formulas of algebra);
- *hubs* i.e. pages that do not necessarily contain information, but (as with Yahoo! pages) have links to pages where the information is stored.

Apart from limiting cases, every page i has both an authority score $au(i)$ and a hub score $h(i)$, that are computed via a mutual recursion. In particular we define the authority of one page as proportional to the sum of the hub scores of the pages pointing to it,

$$au(i) \propto \sum_{j \to i} h(j). \tag{4.1}$$

Similarly, the hub score of one page is proportional to the authority scores of the pages reached from the hub,

$$h(i) \propto \sum_{i \to j} au(j). \tag{4.2}$$

To ensure convergence of the above recursion, a good method is to normalise the values of $h(i)$ and $a(i)$ at every iteration such that $\sum_{i=1}^{n} h(i) = \sum_{i=1}^{n} au(i) = 1$.

HITS algorithm

```
def HITS_algorithm(DG):
    auth={}
    hub={}

    k=1000 #number of steps

    for n in DG.nodes():
        auth[n]=1.0
        hub[n]=1.0

    for k in range(k):
        norm=0.0
        for n in DG.nodes():
            auth[n]=0.0
            for p in DG.predecessors(n):
                auth[n]+=hub[p]
            norm+=auth[n]**2.0
        norm=norm**0.5
        for n in DG.nodes():
            auth[n]=auth[n]/norm
```

```
        norm=0.0
        for n in DG.nodes():
            hub[n]=0.0
            for s in DG.successors(n):
                hub[n]+=auth[s]
            norm+=hub[n]**2.0
        norm=norm**0.5
        for n in DG.nodes():
            hub[n]=hub[n]/norm

        return auth,hub

DG=nx.DiGraph()

DG.add_edges_from([('A','B'),('B','C'),('A','D'), \
                   ('D','B'),('C','D'),('C','A')])

#plot the graph
nx.draw(DG,with_labels=True)

(auth,hub)=HITS_algorithm(DG)

print auth
print hub

#OUTPUT
{'A': 0.31622776601683794, 'C': 0.31622776601683794,
'B': 0.6324555320336759, 'D': 0.6324555320336759}
{'A': 0.7302967433402215, 'C': 0.5477225575051661,
'B': 0.18257418583505539, 'D': 0.36514837167011077}
```

4.3.2 Spectral properties

This method can be (qualitatively, not considering the normalisation problems) described by means of linear algebra. As seen in the first chapter (see Section 1.3), a graph can be equivalently represented by means of a matrix of numbers, that is, with its adjacency matrix, as shown in the graph in Fig. 4.1.

The equation giving rise to the hub score can be written as

$$h(i) \propto \sum_{i \to j} au(j) \to h(i) \propto \sum_{j=1}^{n} a_{ij} au(j) \to \vec{h} \propto A\vec{au} \tag{4.3}$$

and similarly for the authorities we obtain:

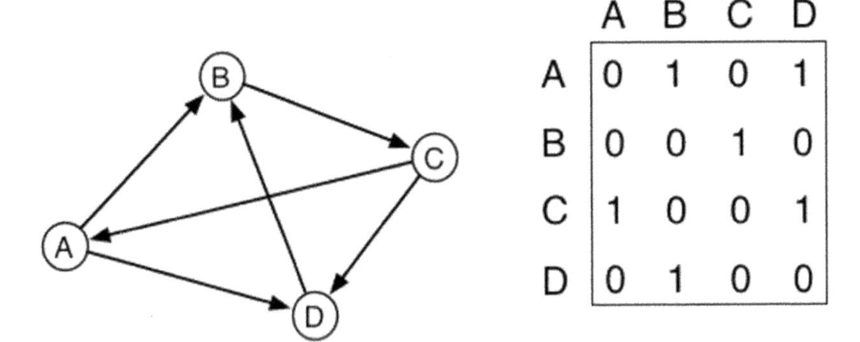

Fig. 4.1 A simple oriented graph with its adjacency matrix.

$$au(i) \propto \sum_{j \to i} h(j) \to au(i) \propto \sum_{j=1}^{n} a^T_{ij} h(j) \to \vec{u} \propto A^T \vec{h}, \tag{4.4}$$

where a^T_{ij} are the elements of the matrix A^T that is the transpose of A (this means that $a^T_{ij} = a_{ji}$.

How to transpose and multiply a matrix

```
def matrix_transpose(M):
    M_out=[]
    for c in range(len(M[0])):
        M_out.append([])
        for r in range(len(M)):
            M_out[c].append(M[r][c])
    return M_out


def matrix_multiplication(M1,M2):
    M_out=[]
    for r in range(len(M1)):
        M_out.append([])
        for j in range(len(M2[0])):
            e=0.0
            for i in range(len(M1[r])):
                e+=M1[r][i]*M2[i][j]
            M_out[r].append(e)
    return M_out
```

```
adjacency_matrix1=[
                  [0,1,0,1],
                  [1,0,1,1],
                  [0,1,0,0]
                  ]

adjacency_matrix2=matrix_transpose(adjacency_matrix1)

print "Transpose adjacency matrix:",adjacency_matrix2

res_mul=matrix_multiplication(adjacency_matrix1,adjacency_matrix2)

print "Matrix multiplication:",res_mul

#OUTPUT
Transpose adjacency matrix: [[0, 1, 0], [1, 0, 1], [0, 1, 0],
[1, 1, 0]]
Matrix multiplication: [[2.0, 1.0, 1.0], [1.0, 3.0, 0.0],
[1.0, 0.0, 1.0]]
```

By combining (4.3) and (4.4) we obtain

$$\vec{h} \propto AA^T\vec{h} = \lambda_h AA^T\vec{h},$$
$$\vec{au} \propto A^TA\vec{au} = \lambda_{au}A^TA\vec{u}. \tag{4.5}$$

That is an eigenvalue problem for the matrices $M \equiv AA^T$ and $M^T \equiv A^TA$.

- M (and therefore its transpose) is real and symmetric, so its eigenvalues are real;
- M is non-negative (i.e. the entries are at least 0 or larger); if we can find a $k > 0$ such that $M^k >> 0$, that is, all of the entries are strictly larger than 0, then M is *primitive*. If M is a primitive matrix:
  * the largest eigenvalue λ of M is positive and of multiplicity 1;
  * every other eigenvalue of M is in modulus strictly less than λ;
  * the largest eigenvalue λ has a corresponding eigenvector with all entries positive.

Being a primitive matrix means in physical terms that the graph defined by the adjacency matrix must have no dangling ends or sinks and that it is possible to reach any page from any starting point. In all of the above hypothesis convergence is ensured.

Principal eigenvalue/vector extraction (power iteration)

```
adjacency_matrix=[
                  [0,1,0,1],
```

```
                   [1,0,1,1],
                   [0,1,0,0],
                   [1,1,0,0]
                   ]
vector=[
        [0.21],
        [0.34],
        [0.52],
        [0.49]
        ]

for i in range(100): #100 iterations is enough for the convergence!
    res=matrix_multiplication(adjacency_matrix,vector)
    norm_sq=0.0
    for r in res:
        norm_sq=norm_sq+r[0]*r[0]
    vector=[]
    for r in res:
         vector.append([r[0]/(norm_sq**0.5)])

print "Maximum eigenvalue (in absolute value):",norm_sq**0.5
print "Eigenvector for the maximum eigenvalue:",vector

#OUTPUT
Maximum eigenvalue (in absolute value): 2.17008648663
Eigenvector for the maximum eigenvalue: [[0.5227207256439814],
[0.6116284573553772], [0.2818451988548684], [0.5227207256439814]]
```

Starting from the data previously downloaded from the laboratory for web algorithmics of the University of Milan, we can now apply the HITS algorithm to the real case of the ".eu" portion of the WWW in 2005. The output will be the top urls and the corresponding auth and hub values.

HITS algorithm for the ".eu" domain in 2005

```
import operator

(auth,hub)=HITS_algorithm(eu_DG_SCC)
sorted_auth = sorted(auth.items(), key=operator.itemgetter(1))
sorted_hub = sorted(hub.items(), key=operator.itemgetter(1))

#top ranking auth
```

```
print "Top 5 auth"
for p in sorted_auth[:5]:
    print dic_nodid_urls[p[0]],p[1]

#top ranking hub
print "\nTop 5 hub"
for p in sorted_hub[:5]:
    print dic_nodid_urls[p[0]],p[1]

#OUTPUT
top 5 auth
http://www.etf.eu.int/WebSite.nsf/... 9.67426387995e-05
http://www.etf.eu.int/website.nsf/Pages/Job... 9.67426387995e-05
http://www.etf.eu.int/WebSite.nsf/(tenders... 9.67426387995e-05
http://europa.eu.int/eures/main.jsp?...LV 9.67426387995e-05
http://europa.eu.int/eures/main.jsp?...DE 9.67426387995e-05

top 5 hub
http://www.etf.eu.int/... 7.65711101121e-07
http://ue.eu.int/cms3_fo/showPage.asp... 7.65711101121e-07
http://ue.eu.int/showPage.asp?id=357... 7.65711101121e-07
http://ue.eu.int/showPage.asp?id=370... 7.65711101121e-07
http://www.europarl.eu.int/interp... 7.65711101121e-07
```

4.3.3 PageRank

HITS is not the only algorithm that assesses the importance of a page by using the spectral properties of the adjacency matrix (or functions of it). Actually, the most successful measure of eigenvector centrality is given by another algorithm, known as PageRank. The idea is similar to that of the HITS algorithm, but now we give only one score to the pages of the web, irrespective of its role as authority or hub. The values of PageRank for the various pages in the graph are given by the eigenvector $\mathbf{r}$, related to the largest eigenvalue λ_1 of the matrix $\mathbf{P}$, given by

$$\mathbf{P} = \alpha\mathbf{N} + (\mathbf{1} - \alpha)\mathbf{E}; \tag{4.6}$$

the weight is taken as $\alpha = 0.85$ in the original paper (Page *et al.*, 1999). $\mathbf{N}$ is the normalised matrix $\mathbf{N} = \mathbf{AK0}^{-1}$ where $\mathbf{A}$ is the adjacency matrix and $\mathbf{K0}^{-1}$ is the diagonal matrix, whose entries on the diagonal are given by the inverse of the out degree, $\left(\mathbf{K0}^{-1}\right)_{ii} = 1/k_i^o$.

This new matrix $\mathbf{P}$ does not differ considerably from the original one $\mathbf{N}$, but has the advantage that (thanks to its irreducibility) its eigenvectors can be computed by a simple iteration procedure Langville and Meyer (2003).

As we have seen for HITS, dealing with a matrix that is not primitive presents a series of problems. The presence of dangling nodes avoids $\mathbf{N}$ being a stochastic matrix

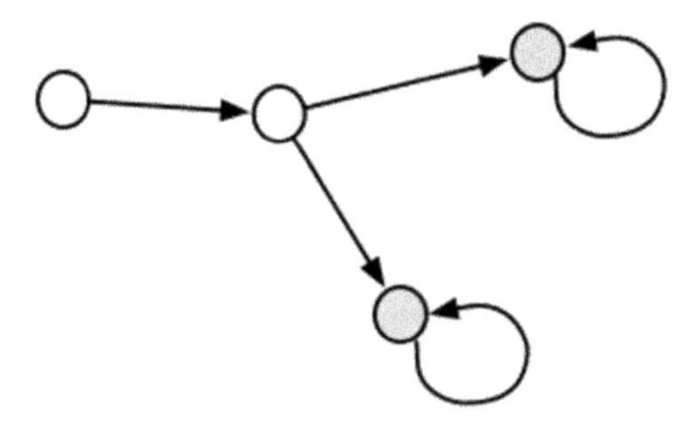

Fig. 4.2 A simple case of reducible matrix.

and therefore gives problems for the existence of the limiting vector $\mathbf{r}^{\infty}$, which is the numerical solution of the equation $\mathbf{r} = \mathbf{rN}$. Even worse, almost certainly, the subgraph represented by $\mathbf{N}$ will be reducible. A reducible stochastic matrix is one for which the underlying chain of transformations is reducible[3]. A reducible chain, is one for which there are states in which the evolution can be trapped. The simplest example of a reducible matrix is that of a page i that has an edge to page j, and this page j has a loop (citing itself) and a link to another page z which again has a loop (an edge to itself) (see Fig. 4.2). Iteration on this set will not produce convergence to a limit vector r^{∞}. When the matrix is *irreducible*, a mathematical theorem (by Perron and Frobenius) ensures that this chain must have a unique and positive stationary vector $\mathbf{r}^{\infty}$ (Perron, 1907; Frobenius, 1912). A physical way to force irreducibility numerically is to destroy the possibility of getting trapped. If you can jump out from a page to a completely random different one (even with small probability), the matrix is irreducible and you can find the eigenvectors $\mathbf{r}$ by iteration. This corresponds to adding to the matrix $\mathbf{N}$ another diagonal matrix $\mathbf{E}$ whose entries e_{ii} are given by $1/n$, where n is the number of vertices in the graph.[4]

Compute the PageRank

```
def pagerank(graph, damping_factor=0.85, max_iterations=100,
min_delta=0.00000001):

    nodes = graph.nodes()
    graph_size = len(nodes)
    if graph_size == 0:
        return {}

    # itialize the page rank dict with 1/N for all nodes
```

[3]In this case the chain is called a Markov chain, since the state at a certain time of evolution depends only upon the state at the previous time step.

[4]In the more recent implementation of PageRank, those entries are actually different from each other, even if they have the same order of magnitude. This is done in order to introduce an ad hoc weight for the different pages.

```
    pagerank = dict.fromkeys(nodes, (1.0-damping_factor)*1.0/ \
                                 graph_size)
    min_value=(1.0-damping_factor)/len(nodes)

    for i in range(max_iterations):
        diff = 0 #total difference compared to last iteraction
        # computes each node PageRank based on inbound links
        for node in nodes:
            rank = min_value
            for referring_page in graph.predecessors(node):
                rank += damping_factor * pagerank[referring_page]/ \
                len(graph.neighbors(referring_page))
            diff += abs(pagerank[node] - rank)
            pagerank[node] = rank

        #stop if PageRank has converged
        if diff < min_delta:
            break

    return pagerank
```

Starting with the following test network, we can apply the Pagerank algorithm with both our code and the NetworkX corresponding function.

PageRank for a test network

```
G=nx.DiGraph()
G.add_edges_from([(1,2),(2,3),(3,4),(3,1),(4,2)])
#plot the network
nx.draw(G)

#our Page Rank algorithm
res_pr=pagerank(G,max_iterations=10000,min_delta=0.00000001, \
                damping_factor=0.85)
print res_pr

#Networkx Pagerank function
print nx.pagerank(G,max_iter=10000)

#OUTPUT
{1: 0.17359086186340225, 2: 0.33260446516778386,
3: 0.3202137953926163, 4: 0.17359086304186191}
```

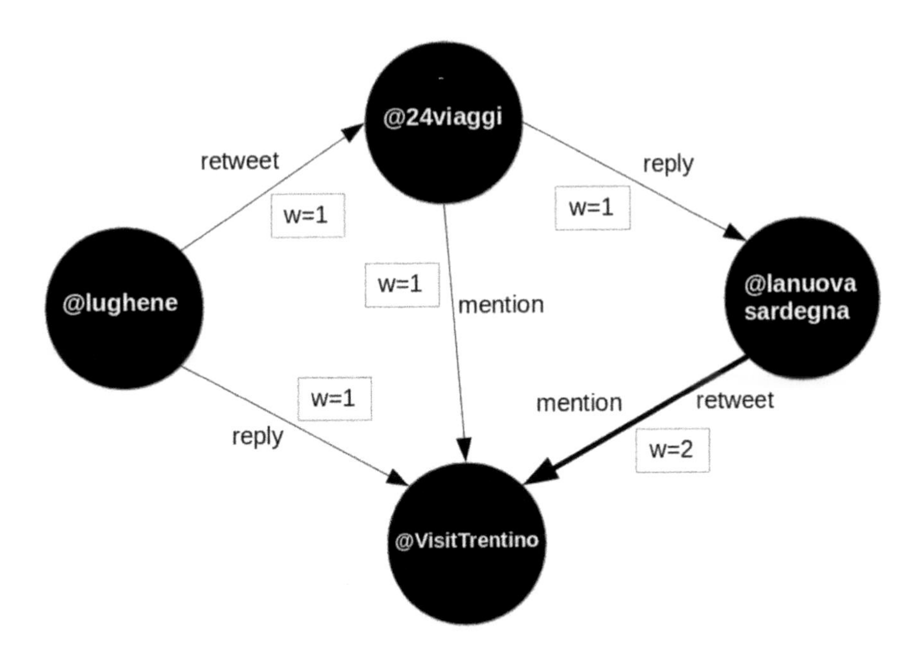

Fig. 4.3 This is the procedure to generate a network starting from a flux of tweets. The nodes are the twitter users and each time one of them mentions, retweets or replies to another user a link is drawn from the first to the second. The weight of a links is the number of citations between the two.

```
{1: 0.17359061775974502, 2: 0.33260554622228633,
3: 0.3202132182582236, 4: 0.17359061775974502}
```

Now we have the opportunity to make sense of this famous algorithm in a real case in the field of social network analysis. At the beginning of this chapter we learnt how to retrieve tweets through the Twitter API. With the "search" method we can get a certain number of tweets and try to uncover the discussion going on related to the search criteria we have imposed. The first step in this process is to map this flux of information in the shape of a network and after that try to measure on it some particular property. More than the so-called structural network of followers and friends, in order to discover the thread of discussions it is useful to generate the network of mentions, retweets, and replies. In this case a link is drawn from a user "A" towards a user "B" if the user "A" mentions user "B" in one of their tweets (see Fig. 4.3).

It is a way of acknowledging someone and giving credit to them, just as happens when a web page relates to another through a hyperlink. This kind of network is able to catch the just-in-time interaction among users about a particular topic, much more than the structural ones. For the present example we will limit ourselves to the mention network only.

Given this procedure we can now apply it to the case of a particular topic and extract in a natural way the thread of the discussion as the clusters that emerge as isolated components of the resulting network (see Fig. 4.4).

Generate and plot the Twitter mention network

```
def generate_network(list_mentions):
    DG=nx.DiGraph()
    for l in list_mentions:
        if len(l)<2: continue
        for n in l[1:]:
             if not DG.has_edge(l[0],n):
                DG.add_edge(l[0],n, weight=1.0 )
        else:
                DG[l[0]][n]['weight']+=1.0
    return DG

#extracting user and mentions for each tweet
res=twitter_connection.search(q='#FutureDecoded', count=5000)
#the first will be the tweer user
list_users={}
list_mentions=[]
for t in res['statuses']:
    list_unique_ids=[]
    print "User Screen Name and Id:",(t[u'user'][u'screen_name'], \
                                      t[u'user'][u'id_str'])
    list_unique_ids.append(t[u'user'][u'id_str'])
    if not list_users.has_key(t[u'user'][u'id_str']):
        list_users[t[u'user'][u'id_str']]=t[u'user'][u'screen_name']
    print "List of Mentions:",
    for m in t[u'entities'][u'user_mentions']:
        if m['id_str']!=t[u'user'][u'id_str']:
                list_unique_ids.append(m['id_str'])
                if not list_users.has_key(m['id_str']):
                    list_users[m['id_str']]=m[u'screen_name']
        print (m[u'screen_name'],m['id_str']),
    print "\r"
    print list_unique_ids
    list_mentions.append(list_unique_ids)
    print "\n"

net_mentions=generate_network(list_mentions)

#plotting the network
```

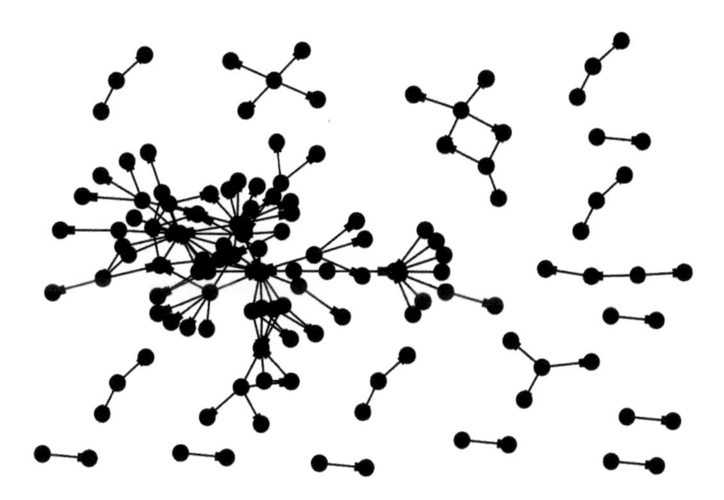

Fig. 4.4 Plot of the mention network arising from Twitter data. The nodes are the Twitter users and the oriented links go from one user to another, mentioned by the first in one of their tweets. In a natural way the threads emerge from the picture as isolated clusters, with a main one dominating the discussion.

```
pos=nx.graphviz_layout(net_mentions,prog='neato')
nx.draw(net_mentions, pos, node_size = 50, node_color='Black')
savefig('./data/hashtag_discussion_thread.png',dpi=600)

#OUTPUT
...
User Screen Name and Id: (u'vincent_salmon', u'294176299')
List of Mentions: (u'TagetikUK', u'3653601856') (u'manuelvellutini',
u'139816639') (u'MicrosoftUK', u'720474368') (u'mspartnersuk',
u'23672986') [u'294176299', u'3653601856', u'139816639',u'720474368',
u'23672986']
...
```

Finally, we can compute the Pagerank on this oriented network getting the most central nodes (users). In this case the simple interpretation is that the top ranking users are very likely the most influential(see also (Perra *et al.*, 2009)), in relation to the selected topic.

Top Pageranks on a Twitter generated network (influencers)

```
pr=nx.pagerank(net_mentions,max_iter=10000)
sorted_pr=sorted(pr.items(), key=operator.itemgetter(1),reverse=True)
#top10 pagerank twitter user from the selected search
for page in sorted_pr[:10]:
    print list_users[page[0]],page[1]

#OUTPUT
microsoftitalia 0.0504950565972
GiacomoFrisoni 0.0480209549756
satyanadella 0.0359068749876
MSFTBusinessUK 0.0350727368583
FabioSantini71 0.032280683387
Microsoft 0.0234986623942
federicadestr 0.022313707312
purassan 0.0177527343108
msdev_ita 0.0109019463377
Fagrossi67 0.0106376988422
```

4.4 Communities and Girvan–Newman algorithm

The concept of communities is not in itself extremely precise, and also therefore methods for determining them in networks are many and refer to slightly different objects. Actually, we can have communities of people corresponding to connected subgraphs of the graph (similar to cliques). On the other hand we can define communities by means of vertices with similar properties (i.e. sharing similar links). In this latter case a community can be determined by a set of vertices which may be totally disconnected. Loosely speaking, when dealing with a large graph we would be interested in a series of vertices and edges that are all somewhat "similar". That is to say we would like to be able to determine some "thematic" subgraphs out of the original (larger) one. Unfortunately, there are various ways of obtaining such a partition of the graph, and *a priori* we cannot ensure that one is better than the others, so it is impossible to tell which method must be used to determine network communities. Imagine the situation of a bipartite graph (i.e. authors on one side and papers on the others). By construction, in such a structure the authors do not have any shared edge, and a community is rather defined by the papers that they are connected to. On the other hand we can transform this graph to an author–author graph where the links connect the persons writing a paper together. In this case the same community is determined by considering which vertices are more closely connected with each other, so that the common edges play a crucial role.

In the following we shall mostly stick to the latest definition of community; showing methods and techniques to determine which subgraphs are composed of vertices which

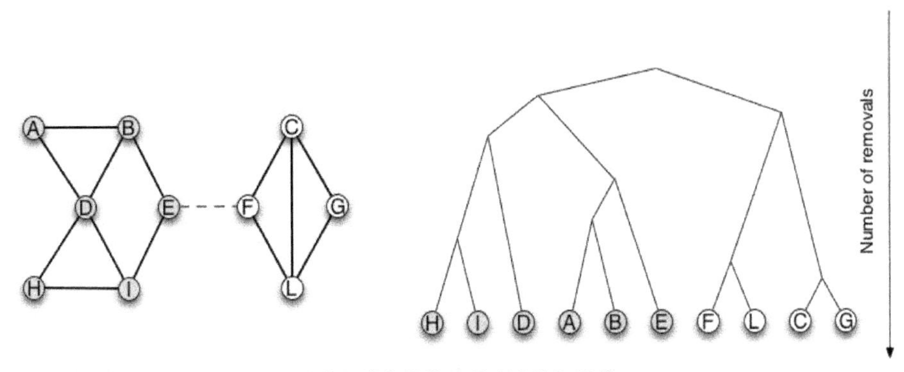

Fig. 4.5 (left) A toy graph to which we applied the GN algorithm. First we compute the edge betweenness and then we cut the edge with the largest value (dashed). Recursively, we compute and again delete all the edges one after another. Whenever the removal of one edge splits the graph, we indicate (right) the edge in bold (i.e. edges **E-F, A-D, G-L, D-I, B-E, A-B, H-I, F-L, C-G**). As a result we obtain the dendrogram on the right.

are closely connected with each other with a strength or density of links larger than the average.

If this is the case in a connected graph, it turns out that communities are denser subgraphs connected by a few links that act as a bridge between them. By cutting out these bridges communities emerge as isolated subgraphs. This is the main idea behind the divisive method of Girvan and Newman (Girvan and Newman, 2002).

4.4.1 Girvan–Newman (GN) algorithm

This method of computing communities, is based on a recursive deletion of edges (Girvan and Newman, 2002; Newman and Girvan, 2004). These edges are not randomly chosen, rather, they are selected for their bridging properties, that is to say they are selected if they connect dense regions and therefore after their removal these dense regions appear as the communities within the system. The quantity chosen for this procedure is the edge betweenness (see Section 3.3). Based on this measure of centrality, one computes the betweenness on all the edges of the graph. We start removing the edge with the largest value then we recompute the edge betweenness and then we delete the one with the largest betweenness among those left. The process is repeated until all the edges are removed. Somewhere during this procedure the structure of communities emerges, however, at different stages we have different sets of communities that vary in both the number of clusters and their size. An example of the procedure is sketched out in Fig. 4.5.

We use a simple toy graph to work out the procedure. We compute the edge betweenness of all the links in the graph. Then we start removing the largest one. Whenever the graph splits into two parts, we keep track of it in the dendrogram. Often, especially at the end of the process, many edges have the same (largest) value of betweeness; in this case we select on of them randomly. This recursive procedure finishes when all the vertices are disconnected. The main problem associated with this way of computing communities, is knowing when to stop the process before we split the network into isolated vertices. Various implementations have been made of this method. For example for large graphs one can compute the betweenness, not by considering all the couples vertices, but just a random selection of the vertices (Tyler *et al.*, 2003). This results in an effective gain in the speed of the algorithm, paying the price of reduced precision.

Code for the GN algorithm

```
G=nx.Graph()
G.add_edges_from([('A','B'),('A','D'),('B','D'),('B','E'),('E','I'),\
                  ('D','I'),('D','H'),('H','I'),('E','F'),('F','C'),\
                  ('F','L'),('C','L'),('C','G'),('G','L')])

pos=nx.graphviz_layout(G,prog='neato')

nx.draw(G, pos,with_labels=True)

#NOTE: THE ORDER OF EDGES IS DIFFERENT FOR THE FACT THAT MANY
#OF THEM HAVE THE SAME BETWEENNESS VALUE...

sorted_bc=[1]
actual_number_components=1
while not sorted_bc==[]:
    d_edge=nx.edge_betweenness_centrality(G)
    sorted_bc = sorted(d_edge.items(), key=operator.itemgetter(1))
    e=sorted_bc.pop()
    print "deleting edge:", e[0],
    G.remove_edge(*e[0])
    num_comp=nx.number_connected_components(G)
    print "...we have now ",num_comp," components"
    if num_comp>actual_number_components:
        actual_number_components=num_comp

#OUTPUT
deleting edge: ('E', 'F') ...we have now  2  components
deleting edge: ('B', 'E') ...we have now  2  components
deleting edge: ('D', 'I') ...we have now  2  components
deleting edge: ('D', 'H') ...we have now  3  components
deleting edge: ('I', 'H') ...we have now  4  components
deleting edge: ('F', 'L') ...we have now  4  components
deleting edge: ('C', 'F') ...we have now  5  components
```

```
deleting edge: ('B', 'D') ...we have now  5  components
deleting edge: ('A', 'B') ...we have now  6  components
deleting edge: ('G', 'L') ...we have now  6  components
deleting edge: ('C', 'G') ...we have now  7  components
deleting edge: ('A', 'D') ...we have now  8  components
deleting edge: ('C', 'L') ...we have now  9  components
deleting edge: ('E', 'I') ...we have now  10  components
```

4.5 Modularity

The whole idea behind the GN algorithm is that the communities are the set of subgraphs that have a link density larger than "expected" (for a random graph of the same size and measure). By cutting bridging edges we isolate such communities and we are able to determine them quantitatively. Since this process does not tell us when one division is better than another, we need a quantity for assessing how good the division is and therefore when we should stop. This quantity is called *modularity* (Newman, 2006) and it assigns a score to any division in clusters one obtains from a given graph. The steps we need to take in order to define this quantity are as follows:

- the starting point is to consider a partition of the graph into g subgraphs;
- if the partition is good most of the edges will be inside the subgraphs and few will connect them;
- we then define a $g \times g$ matrix E whose entries e_{ij} give the fraction of edges that in the original graph connect subgraph i to subgraph j;
- the actual fraction of edges in subgraph i is given by element e_{ii};
- the quantity $f_i = \sum_{j=1,g} e_{ij}$ gives the probability that an end-vertex of a randomly extracted edge is in subgraph i ($i \in 1, ..., g$);
- in the absence of correlations the probability that an edge belongs to subgraph i is f_i^2.

We can now define the modularity Q of a given partition by considering the actual distribution of edges in the partition, with respect to the one we have for a random case, i.e.

$$Q = \sum_{i=1}^{g} e_{ii} - f_i^2, \tag{4.7}$$

which represents a measure of the validity of a certain partition of the graph. In the limit case where we have a random series of communities, the edges can be with the same probability in the same subgraph i or between two different subgraphs i, j. In this case $e_{ii} = f_i^2$ and $Q = 0$. If the division into subgraphs is appropriate, then the actual fraction of internal edges e_{ii} is larger than the estimate f_i^2, and the modularity is larger than zero. Surprisingly, random graphs (which, as we shall see, are graphs obtained by randomly drawing edges between vertices) can present partitions with large modularity (Guimerà *et al.*, 2004). In random networks of finite size it is possible to find a partition which not only has a nonzero value of modularity, but even quite high values. For example, a network of 128 nodes and 1024 edges has a maximum modularity of 0.208. While on average we expect a null modularity for a random graph, this does not exclude that by careful choice we can obtain a different result. This suggests that those networks that seem to have no structure actually exhibit community structure due to fluctuations.

Community detection with the Karate Club network (See Fig. 4.6)

```
import community

G=nx.read_edgelist("./data/karate.dat")

#first compute the best partition
partition = community.best_partition(G)

#plot the network
size = float(len(set(partition.values())))
pos = nx.spring_layout(G)
count = 0.
plt.axis('off')
for com in set(partition.values()) :
    count = count + 1.
    list_nodes = [nodes for nodes in partition.keys() \
                 if partition[nodes] == com]
    nx.draw_networkx_nodes(G, pos, list_nodes, node_size = 300, \
                           node_color = str(count / size))
    nx.draw_networkx_labels(G,pos)

nx.draw_networkx_edges(G,pos, alpha=0.5,width=1)
savefig('./data/karate_community.png',dpi=600)
```

We can perform the same analysis on the Sardinian Wikipedia with the aim of extracting the relevant communities. The first thing to do is to load the network and define the dictionary that associates the Wikipedia node_ids with the page titles.

Community detection for the scwiki web graph

```
#load the directed and undirected version og the scwiki graph
scwiki_pagelinks_net_dir=nx.read_edgelist \
("./data/scwiki_edgelist.dat",create_using=nx.DiGraph())
scwiki_pagelinks_net=nx.read_edgelist("./data/scwiki_edgelist.dat")
```

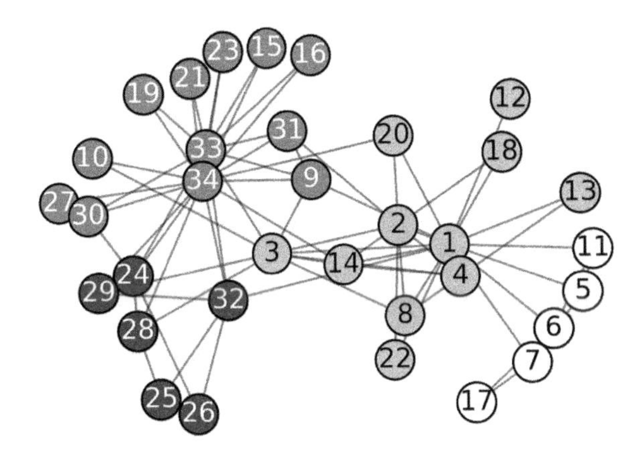

Fig. 4.6 The Karate Club social network after optimisation of the modularity function.

```
#load the page titles
diz_titles={}
file_titles=open("./data/scwiki_page_titles.dat",'r')
while True:
    next_line=file_titles.readline()
    if not next_line:
        break
    print next_line.split()[0],next_line.split()[1]
    diz_titles[next_line.split()[0]]=next_line.split()[1]

file_titles.close()

#OUTPUT
...
4311 Lod
7157 Logos_Bascios
13786 Logroo
8490 Logudoresu
4548 Logudoro
4825 Logusantu
4900 Loiri-Poltu_Santu_Paolu
...
```

The problem in plotting this network is that it comprises almost 10,000 nodes. To overcome this problem we generate a representative network in which each node is a community (we consider just the first nine with more than 200 nodes), with size proportional to the number of nodes in the corresponding community and edge weight proportional to the number of edges between each pair of communities (we cut the link below the threshold weight 100). The representative node is chosen according to the Pagerank inside the corresponding community. So as a first step we generate the representative nodes and print the association with the page title of the scwiki Wikipedia page and the edges with the appropriate weights. The output will be the community id, the number of nodes in it, the page title, and the corresponding PageRank.

Generate and optimise the representative network of the community structure

```
#optimization
partition = community.best_partition(scwiki_pagelinks_net)

#Generate representative nodes of the community structure
community_structure=nx.Graph()
diz_communities={}
diz_node_labels={}
diz_node_sizes={}
max_node_size=0
for com in set(partition.values()) :
    diz_communities[com] = [nodes for nodes in partition.keys() \
                            if partition[nodes] == com]
    if len(diz_communities[com])>=200:
        if max_node_size<len(diz_communities[com]):
            max_node_size=len(diz_communities[com])
        print "community",com,len(diz_communities[com]),
        sub_scwiki_dir = scwiki_pagelinks_net_dir.subgraph \
        (diz_communities[com])
        res_pr=nx.pagerank(sub_scwiki_dir,max_iter=10000)
        sorted_pr=sorted(res_pr.items(), key=operator.itemgetter \
                         (1),reverse=True)
        print diz_titles[sorted_pr[0][0]],sorted_pr[0][1]
        community_structure.add_node(com)
        diz_node_labels[com]=diz_titles[sorted_pr[0][0]]
        diz_node_sizes[com]=len(diz_communities[com])

#Generate edge weights according to the number of links
#among communities
max_edge_weight=0.0
for i1 in range(community_structure.number_of_nodes()-1):
```

```
    for i2 in range(i1+1,community_structure.number_of_nodes()):
        wweight=0.0
        for n1 in diz_communities[community_structure.nodes()[i1]]:
            for n2 in diz_communities[community_structure.nodes() \
                                        [i2]]:
                if scwiki_pagelinks_net.has_edge(n1,n2):
                    wweight=wweight+1.0
        if wweight>100.0:
            if max_edge_weight<wweight:
                max_edge_weight=wweight
            community_structure.add_edge(community_structure. \
            nodes()[i1],community_structure.nodes()[i2], \
                                             weight=wweight)

#OUTPUT

community 0 2012 Logudoresu 0.0507648029074
community 1 1812 Wikipedia 0.0498390890911
community 2 861 Classificatzione_sientfica 0.0358451735252
community 3 795 Babel 0.0296037336974
community 4 662 Sardigna 0.0403623149927
community 5 393 Nugoresu 0.0781774191072
community 6 1939 Limba_Sarda_Comuna 0.0219626233778
community 8 201 Rabascius 0.053309971387
community 22 223 Casteddu 0.0197102893607
```

The final plot will reveal the hierarchy of the community structure, the relative sizes of nodes/communities and edges (see Fig. 4.7). This result shows that the most important nodes in the communities are related to territorial locations and language classifications, topics that are known to be relevant to the Sardinian culture.

Plotting the representative network of the community structure

```
pos=nx.graphviz_layout(community_structure,prog='circo')
node_size_factor=2000.0
edge_weight_factor=10.0

plt.axis('off')

for n in community_structure.nodes():
    nx.draw_networkx_nodes(community_structure, pos, [n], node_size\
                           = node_size_factor*diz_node_sizes[n]/ \
                           max_node_size, node_color='Black')
```

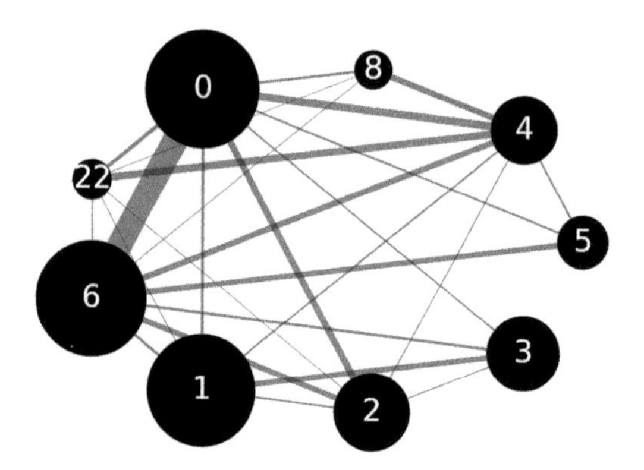

Fig. 4.7 Network representation of the community structure of the Sardinian Wikipedia scwiki. We selected the main communities (the first nine, with more than 200 nodes) the size of each node being proportional to the number of nodes in each community. The edge width is instead proportional to the number of edges between each pair of communities.

```
    nx.draw_networkx_labels(community_structure,pos, font_color= \
                            'White',axis='off')

for e in community_structure.edges():
    nx.draw_networkx_edges(community_structure,pos,[e],alpha=0.5, \
                           width=edge_weight_factor* \
                           community_structure[e[0]][e[1]]['weight']\
                           /max_edge_weight)
```

5 Financial Networks

5.1 Introduction

Activity in the stock markets, has always attracted a great deal of interest from not only investors, but also scientists. Both for different (or maybe the same) reasons, have wanted to discover regularity in price fluctuations. The theory of random walks (Bachelier, 1900) and fractals theory (Mandelbrot, 1963) both originated from studies on pricing of commodities. Continuous storage of a variety of data i.e. numbers of transactions, pricing, numbers of bids and asks for all traded stocks worldwide constantly produces one of the largest datasets available to researchers. As mentioned, this discipline has attracted over time the interest of investors convinced that it would in principle be possible to predict the future behaviour by inspecting the past history. It is noteworthy that since the market is not totally isolated, if many believe that the price will go up, then the price will effectively go up (self-fulfilling prophecy). This creates an interesting feedback between observer and system observed. What can be considered to be certain is that we have clear evidence of correlations in the form of self-affinities in price history, with the presence of characteristic roughness exponents. The study of time series is just one of the ways in which we can study quantitatively economic and financial networks. Another approach that is particularly fruitful is to describe the various connections between financial institutions in the form of a network. The structure obtained is particularly complex, since an edge (or various kinds of edges) can represent lending, exposure, insurance, credit default swaps (CDS), ownership, interlock in the board etc. The aim of this chapter is to provide the reader with the main quantitative instruments to describe these systems. Codes, data and/or links for this chapter are available from **http://book.complexnetworks.net**.

5.2 Data from Yahoo! Finance

Financial data are very difficult to collect, essentially due to disclosure problems, but also because of the absence of specific policy regulations on certain kinds of transactions; also most of the data are not available in an aggregated form. Nevertheless, after the financial crisis which started with sub-prime mortgages in 2008, it became clear to a variety of policy regulators and control organisations, that the complexity of the financial structure and our poor knowledge of it had been one of the causes of the turndown in the economy. From that moment a series of international organisations and companies started collecting and making available various data, unfortunately not always accessible to scientists. The set of data we present here has been downloaded from the Yahoo! Finance web service, which offers daily historical data for the closure prices of stock traded in various markets. In the following we present how to interact with the service in order to get the relevant data we need to explore the correlations between stocks for companies present in the NYSE (New York Stock Exchange) index.

Connecting with the Yahoo! Finance service

```
import yahoo_finance as yf

yahoo = yf.Share('YHOO')
d=yahoo.get_historical('2014-05-19', '2014-05-20')
print "A week of stock daily quotations:"
for e in d:
    print e
print "Info about the company:",yahoo.get_info()
print "Market capitalization in dollars:",yahoo.get_market_cap()

OUTPUT
A week of stock daily quotations:
{'Volume': '18596700', 'Symbol': 'YHOO', 'Adj_Close': '33.869999',
'High': '34.470001', 'Low': '33.669998', 'Date': '2014-05-20',
'Close': '33.869999', 'Open': '33.990002'}
{'Volume': '14845700', 'Symbol': 'YHOO', 'Adj_Close': '33.889999',
'High': '33.990002', 'Low': '33.279999', 'Date': '2014-05-19',
'Close': '33.889999', 'Open': '33.41'}

Info about the company: {'start': '1996-04-12', 'symbol': 'YHOO',
'end': '2015-06-17', 'CompanyName': None}

Market capitalization in dollars: 38.13B
```

The historical data from Yahoo! Finance presents information about the volume of stocks transacted, the highest, the lowest, the opening, and the closing values, as well as an adjusted closing value that provides the closing price (on the requested day, week, or month for any stock) adjusted for all applicable splits and dividend distributions. Starting from this data we can easily compute the total transaction volume for the day as the product of the number of shares exchanged and the adjusted closing value.

Transaction volumes computation and plotting (see Fig. 5.1)

```
d=yahoo.get_historical('2014-01-01', '2014-12-31')
V = []
for s in d:
    print s['Date'],float(s['Volume'])*float(s['Adj_Close'])
```

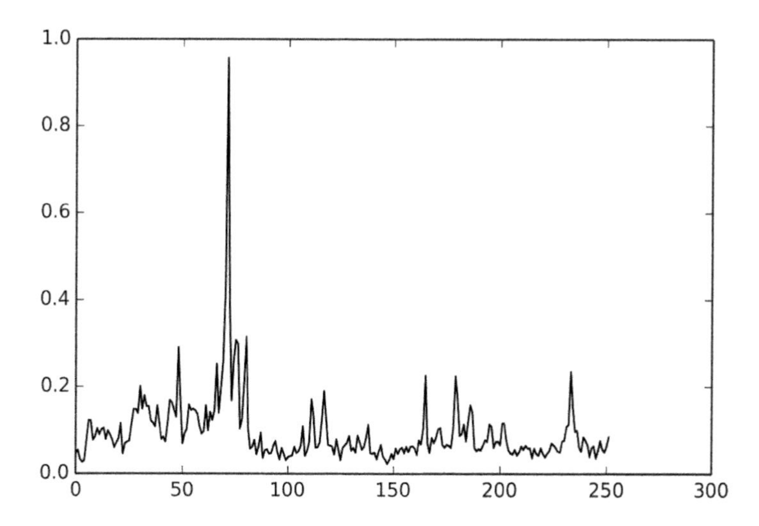

Fig. 5.1 One year of stock quotations for the Yahoo share from January 01, 2014, up to December 31, 2014. The values are in 10^{10}\$ (US Dollars).

```
    V.append(float(s['Volume'])*float(s['Adj_Close']))

plot(V)
savefig('yahoo_volume.png')

#OUTPUT
2014-12-31 469995531.39
2014-12-30 548233280.704
2014-12-29 334735978.375
2014-12-26 262930947.17
2014-12-24 301970246.924
2014-12-23 776010280.0
2014-12-22 1228679313.04
2014-12-19 1226727000.11
```

We could also get all information related to shares present in the NYSE index querying the Yahoo! Finance service, but here we will follow a mixed and hopefully simpler approach. We will retrieve the sector and industry from a web page where it is possible to download a CSV file with all of this information (http://www.nasdaq.com/screening/company-list.aspx), while the actual market capitalisation will be obtained from the Yahoo service. Only companies with a capitalisation greater than 50 billion dollars will be considered in our analysis.

Get stock labels, sector, and industries

```
#this code will take approximative 1 hour to retrieve the data
#depending on the internet connection
#if you want to skip this procedure just uncomment
#the following lines
#import sys
#f=open("./data/list_stocks_50B_6_may_2016.txt",'r')
#list_stocks=[]
#while True:
#    next_line=f.readline()
#    if not next_line: break
#    list_stocks.append(tuple(next_line.split('\t')[:-1]))
#f.close()
#sys.exit()

import time

hfile=open("./data/companylist.csv",'r')
#we choose to get only companies with a market capitalisation
#greater than 50B$
cap_threshold=50.0

list_stocks=[]
nextline=hfile.readline()
while True:
    nextline=hfile.readline()
    if not nextline:
        break
    line=nextline.split(',')
    sym=line[0][1:-1]
    share = yf.Share(sym)
    y_market_cap=share.get_market_cap()
    if not y_market_cap: continue
    #we will exclude stocks with char ^ that will
    #give errors in the query process
    if y_market_cap[-1]=='B' and float(y_market_cap \
        [:-1])>cap_threshold and line[0].find('^')==-1:
        print sym,y_market_cap
        list_stocks.append((line[0][1:-1],line[1][1:-1],\
                            line[5][1:-1],line[6][1:-1]))
    time.sleep(1)
```

```
hfile.close()

print list_stocks[0]

OUTPUT
MMM 99.27B
ABB 50.54B
ABT 72.17B
ABBV 106.37B
ACN 60.56B
AEB 50.61B
..........

('MMM', '3M Company', 'Health Care', 'Medical/Dental Instruments')
```

When we need to plot using specific colour codes for companies in the plot, and we need specific dictionaries to handle companies, colours, and sectors.

Generate dictionaries for companies, sectors, and colours

```
diz_sectors={}
for s in list_stocks:
    diz_sectors[s[0]]=s[2]

list_ranking=[]
for s in set(diz_sectors.values()):
    list_ranking.append((diz_sectors.values().count(s),s))

list_ranking.sort(reverse=True)

#list_colors=['red','green','blue','black''cyan','magenta','yellow']
list_colors=['0.0', '0.2', '0.4', '0.6','0.7', '0.8', '0.9']

#'white' is an extra color for 'n/a' and 'other' sectors

diz_colors={}

#association color and more represented sectors
for s in list_ranking:
    if s[1]=='n/a':
        diz_colors[s[1]]='white'
        continue
    if list_colors==[]:
```

```
        diz_colors[s[1]]='white'
        continue
    diz_colors[s[1]]=list_colors.pop(0)
```

5.3 Prices time series

The time series of a stock price is a typical quantity that investors (right or wrong) use when considering their investments (we do not comment here whether they are right or not in doing so). This field of finance is particularly awkward to study since phenomena observed are heavily affected by our actions. As already mentioned, if all investors of a stock suddenly believe that a signal in a stock price time series is indicating impending bankruptcy, all of them will sell the stock causing the bankruptcy for real (a typical case of a self-fulfilling prophecy).

Retrieving historical data

```
start_period='2013-05-01'
end_period='2014-05-31'
diz_comp={}
for s in list_stocks:
    print s[0]
    stock = yf.Share(s[0])
    diz_comp[s[0]]=stock.get_historical(start_period, end_period)

#create dictionaries of time series for each company
diz_historical={}
for k in diz_comp.keys():
    if diz_comp[k]==[]: continue
    diz_historical[k]={}
    for e in diz_comp[k]:
        diz_historical[k][e['Date']]=e['Close']

for k in diz_historical.keys():
    print k,len(diz_historical[k])
```

In the (strong) hypothesis that, in calm periods, various psychological effects cancel out, investors study the statistical properties of the time series, trying to spot regularities that could anticipate the future behaviour of the price. While the link between past and future performance has never been demonstrated, there is nevertheless a certain consensus that "on average" this information is valuable to the investors. In particular the return and the volatility are considered the most important indicators. Given a time interval Δt, let us consider an asset price at the beginning $p(t_0)$ and at the end $p(t_0 + \Delta t)$. We define the proportional return of the investment in the period

Δt as

$$r(\Delta t) = \frac{p(t_0 + \Delta t) - p(t_0)}{p(t_0)}. \tag{5.1}$$

Here we assumed investment in only a certain number of one type of stock, so that we can use the price to determine costs and gains. The above equation in the limit ($\Delta t \to 0$) can be written as $r(t) \simeq \frac{d \ln (p(t))}{dt}$.

This expression passing to discrete time steps takes the following form:

$$r = \ln p(t_0 + \Delta t) - \ln p(t_0). \tag{5.2}$$

Return of prices

```
reference_company='ABEV'
diz_returns={}
d=diz_historical[reference_company].keys()
d.sort()
print len(d),d

for c in diz_historical.keys():
    #check if the company has the whole set of dates
    if len(diz_historical[c].keys())<len(d): continue
    diz_returns[c]={}
    for i in range(1,len(d)):
        #price returns
        diz_returns[c][d[i]]=math.log( \
        float(diz_historical[c][d[i]])) \
        -math.log(float(diz_historical[c][d[i-1]]))

print diz_returns[reference_company]
```

Among the various definitions of volatility σ, the simplest is the standard deviation of the value of prices $p(t)$.

Basic statistics and the correlation coefficient

```
#mean
def mean(X):
    m=0.0
    for i in X:
        m=m+i
    return m/len(X)

#covariance
```

```
def covariance(X,Y):
    c=0.0
    m_X=mean(X)
    m_Y=mean(Y)
    for i in range(len(X)):
        c=c+(X[i]-m_X)*(Y[i]-m_Y)
    return c/len(X)

#pearson correlation coefficient
def pearson(X,Y):
    return covariance(X,Y)/(covariance(X,X)**0.5 * \
                              covariance(Y,Y)**0.5)
```

5.4 Correlation of prices

In the same spirit, correlations in time series (or more simply comovements) are also considered to be extremely valuable. The idea is that every investor has precise knowledge of the market (highly unrealistic (Greenwald, Bruce and Stiglitz, 1993)) and since (s)he is perfectly rational (another strong assumption), (s)he wants to maximise the return and at the same time minimise the risk of their investments. This is obtained by choosing the proportion of the investments among all the assets present in the market (considered "complete") and by essentially building a portfolio of all the different assets. All these concepts have been formalised in the "Theory of portfolio"(Markowitz, 1952) and constitute the basis of operation for professional investors.

Coming back to the real market, if two or more assets have a past history of common behaviour (i.e. they both go up or down at the same time) we can measure a correlation between their price evolution as given by these "comovements". Of course there is no proof that the presence of such a correlation in the past is also a good proxy of its presence in the future. On the other hand, it may very well be that two assets belonging to the same industrial sector have similar behaviour. For example when few firms producing high technology objects (such as computers) go up in the market, it is true that all the assets linked to that technology go up as well. Another example could be when the first asset owns a part of the second, so that a movement (up or down) of one causes the same movement in the second. In these exempla if the correlation in the price is caused by a "hidden link" between the assets, it is fair to assume that we shall also have price correlation in the future. This is important to know when we build our portfolio. Having many assets all behaving the same is like putting all our eggs in a single basket, thereby reducing the risk protection. In a market of say 1000 assets, correlations are of the order of millions (since for N assets, the independent correlations we need to check are of the order of N^2) a number typically too large to allow eye-inspection analysis. Therefore filtering of information is necessary before proceeding to any choice of investment.

The crucial variable is the daily closure price $r_i(t)$ of company i on day t. From that, one can consider all the possible pairs of companies and compute the correlation between the respective price returns. Two price stocks are correlated if they vary in a similar way. In other words companies i and j are correlated when the price of stock i increases if the price of stock j also increases. To quantify such a relation, we compute the correlation $\rho_{ij}(\Delta t)$ between the price returns over a time Δt. Correlation is computed by means of

$$\rho_{ij}(\Delta t) = \frac{\langle r_i r_j \rangle - \langle r_i \rangle \langle r_j \rangle}{\sqrt{(\langle r_i^2 \rangle - \langle r_i \rangle^2)(\langle r_j^2 \rangle - \langle r_j \rangle^2)}}. \tag{5.3}$$

By definition, $\rho_{ij}(\Delta t)$ can vary from -1 (when stocks i and j are completely anti-correlated) to 1 (when stocks i and j are completely correlated). In between there is another important situation: when $\rho_{ij}(\Delta t) = 0$ the two stocks i and j are uncorrelated. Given its meaning, the matrix of the correlation coefficient is symmetric with a value of $\rho_{ij}(\Delta t) = 1$ along the main diagonal (autocorrelation).

Correlation of price returns

```
def stocks_corr_coeff(h1,h2):
    l1=[]
    l2=[]
    intersec_dates=set(h1.keys()).intersection(set(h2.keys()))
    for d in intersec_dates:
        l1.append(float(h1[d]))
        l2.append(float(h2[d]))
    return pearson(l1,l2)

#correlation with the same company has to be 1!
print stocks_corr_coeff(diz_returns[reference_company], \
                        diz_returns[reference_company])

OUTPUT
1.0
```

5.5 Minimal spanning trees

We have already seen that both Traceroute paths and food webs can be represented in the form of trees. Trees are economical graphs in the sense that they connect a fixed number of vertices through the minimal number of edges. One can further reduce the number of links in a tree, by dividing it into two parts and creating more, smaller sub-clusters. A set of disjoint (sub-)trees is called (intuitively) a forest.

Given this "economical" feature, it is hardly a surprise that they are very frequently used in complex systems. For similar reasons, trees are also often used to investigate network structure, as in the case of the breadth first search algorithms and/or as in this case, to filter the information present in a complete graph. More generally, trees are perfect for classifying information. In the case of botany or zoology, this is very easy and is the basis of taxonomic trees. We start from species and we cluster them according to their morphology. Classes of species can be clustered in the same way. Step by step we form a tree composed of different layers.

Using the correlation values previously defined we obtain a set of $n \times (n-1)/2$ numbers characterising the similarity of any of the n stocks with respect to all the other $n-1$ stocks. This set of numbers forms a complete graph with different edge strengths given by the correlation value. At this point we use trees to filter the information reducing the density of the graph. To every entry of the above-defined correlation matrix we can associate a metric distance between any pair of stocks by defining

$$d_{i,j}(\Delta t) = \sqrt{2(1 - \rho_{ij}(\Delta t))}. \tag{5.4}$$

With this choice, $d_{i,j}(\Delta t)$ fulfils the three axioms of a metric distance:

- $d_{i,j}(\Delta t) = 0$ if and only if $i = j$;
- $d_{i,j}(\Delta t) = d_{j,i}(\Delta t) \forall i, j$;
- $d_{i,j}(\Delta t) \leq d_{i,k}(\Delta t) + d_{k,j}(\Delta t) \forall i, j, k$.

The distance matrix $D(\Delta t)$ is then used to determine the MST connecting the n stocks (Gower, 1966; Mantegna, 1999).

Building the network with the metric distance

```
import math
import networkx as nx

corr_network=nx.Graph()

num_companies=len(diz_returns.keys())
for i1 in range(num_companies-1):
    for i2 in range(i1+1,num_companies):
        stock1=diz_returns.keys()[i1]
        stock2=diz_returns.keys()[i2]
        #metric distance
        metric_distance=math.sqrt(2*(1.0-stocks_corr_coeff\
                (diz_returns[stock1],diz_returns[stock2])))
        #building the network
        corr_network.add_edge(stock1, stock2, weight=metric_distance)

print "number of nodes:",corr_network.number_of_nodes()
print "number of edges:",corr_network.number_of_edges()
```

The method for constructing the MST linking N objects is known in multivariate analysis as the "nearest neighbour single linkage cluster algorithm" (Mardia *et al.*, 1979). The idea is to consider the above-defined distance (5.4) between two vertices as the weight of the link connecting them. At this point we keep only the strongest correlations or the shortest distances. To filter among the $\simeq n^2$ links we first rank all the edges, then we start from the vertices which are nearest and we keep adding new vertices by following the rank of the edges, discarding all the links that would form a cycle (in this way, by construction, the graph is acyclic, i.e. a tree). Finally, we stop when all the vertices are drawn (in this way the tree is spanning). Schematising:

1. rank a couple of vertices (stocks) from the nearest to the farthest
2. draw the first edge from this rank
3. continue in the rank
4. if the new edge does not close a cycle draw it
5. go to point 3
6. stop when all the vertices have been drawn.

Minimal spanning tree (Prim's algorithm)

```
tree_seed=reference_company
N_new=[]
E_new=[]
N_new.append(tree_seed)
while len(N_new)<corr_network.number_of_nodes():
    min_weight=10000000.0
    for n in N_new:
        for n_adj in corr_network.neighbors(n):
            if not n_adj in N_new:
                if corr_network[n][n_adj]['weight']<min_weight:
                    min_weight=corr_network[n][n_adj]['weight']
                    min_weight_edge=(n,n_adj)
                    n_adj_ext=n_adj
    E_new.append(min_weight_edge)
    N_new.append(n_adj_ext)

#generate the tree from the edge list
tree_graph=nx.Graph()
tree_graph.add_edges_from(E_new)

#setting the color attributes for the network nodes
for n in tree_graph.nodes():
    tree_graph.node[n]['color']=diz_colors[diz_sectors[n]]
```

Printing the financial minimum spanning tree (see Fig. 5.2)

```
pos=nx.graphviz_layout(tree_graph,prog='neato', \
```

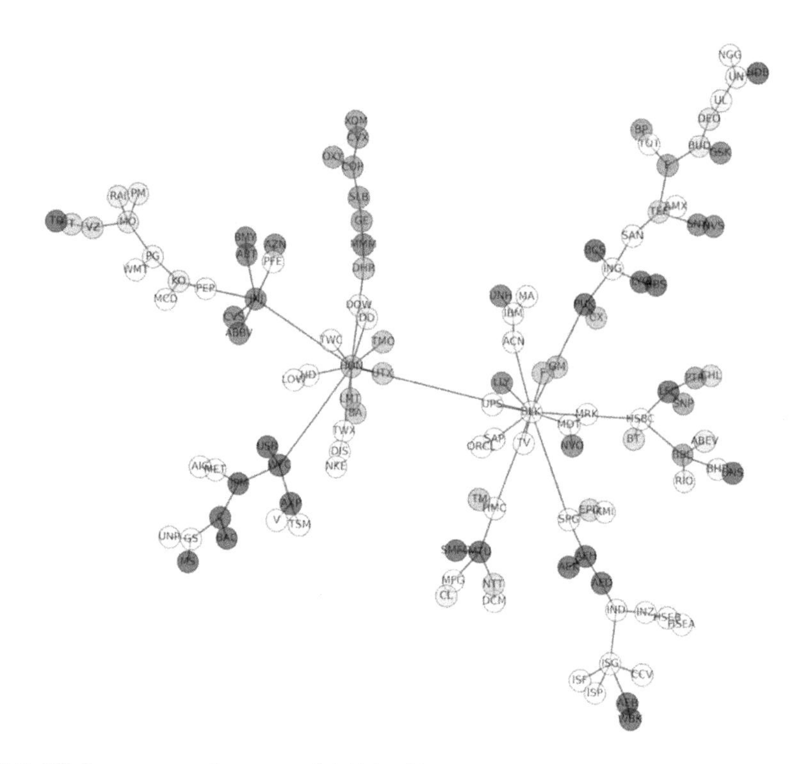

Fig. 5.2 Minimum spanning tree of 141 highly capitalised stocks traded in the US equity markets (NYSE). The filtering procedure has been obtained by considering the correlation coefficient of stock returns time series computed at a one trading day time horizon (6 h and 30 min). Each circle represents a stock labelled by its tick symbol. The minimum spanning tree presents a large amount of stocks having a single link and some stocks having several links. Some of these stocks act as the "hub" of a local cluster.

```
                        args='-Gmodel=subset -Gratio=fill')

figure(figsize=(20,20))
nx.draw_networkx_edges(tree_graph,pos,width=2, \
                        edge_color='black', alpha=0.5, style="solid")
nx.draw_networkx_labels(tree_graph,pos)
for n in tree_graph.nodes():
    nx.draw_networkx_nodes(tree_graph, pos, [n], node_size = 600, \
    alpha=0.5, node_color = tree_graph.node[n]['color'], \
    with_labels=True)

axis('off')

savefig('./data/MST_50B_new.png',dpi=600)
```

6
Modelling

6.1 Introduction

As shown in previous chapters the application of network theory is really multidisciplinary and faces similar problems across the various topics. In particular it is always important to be able to reproduce "synthetically" in a computer the system we are studying. This is the basis of the art of modelling, which allows various and important applications. We can use models to predict the outcome of an experiment, or to understand the basic principles shaping the evolution of a given phenomenon. Seldom, is the mathematical form of the model simple enough to allow an analytical derivation of system evolution. More often this evolution is determined by means of computer simulations. In any case, it is always validation with data from experiments that allows us to determine if the model hypothesis is right or not. In some cases the range of variability of the parameters in the system we are considering can be rather wide; for example the density of an interbank network can vary a lot. In these situations it might happen that different (simple) models hold according to the limiting situations. The aim of this chapter is to present the most used models in the field of complex networks, to illustrate the basic principles on which they are based (sometimes inspired by similar situations in different fields), and to provide the code for them. Given the nature of this chapter, no data repositories are described here, but as usual codes related to this chapter can be downloaded from **http://book.complexnetworks.net**.

6.2 Exponential growth, chains, and random graph

Simple statistical models have always been used in order to determine the evolution of physical systems. In the case of ecology, the first simple model traces back to the Middle Ages with Fibonacci. Later on, more and more refined models were made, trying to detect several quantities across different datasets. This was done by looking at theratio prey/predators (Cohen, 1977) or proportion of top, basal, intermediate species (Pimm *et al.*, 1991; Martinez, 1991). Unfortunately, irrespective of the enormous amounts of work done, scientists are always facing the problem of the relatively small size (and measure) of samples collected. This is a crucial point since if by chance we are missing one or more predations in a food web, we can face dramatic changes of the quantities of interest in a small system. More generally, scientists have tried to model the ecology of species either with or without taking into explicit consideration the predations between species. Models differ for different time scales at which the analysis is done; those who are not considering extinction and mutation explicitly focus on the population of individuals; others look at a much longer scale and try to measure the effects of

t=	1	2	3	4	5	6	7	8	9
n=	1	1	2	3	5	8	13	21	34

Table 6.1 $n(t)$ counts the pairs of rabbits. The formula is $n(t) = n(t-1) + n(t-2)$. That is, at every time step we have all the rabbits we had before, plus a number of sons that come from the rabbit pairs present two time steps before (they need one time step to arrive at sexual maturity.

Darwinian selection in species evolution. Most of the ideas introduced and models used are widely adopted in other contexts, such as, for example, population dynamics that is widely used in *agent-based modelling.*

6.2.1 Static models

One species: Fibonacci sequence. The simplest assumption of a food web, on a short time scale where mutation does not happen is when we deal with one species and unlimited resources. This is a highly unlikely assumption both in ecology and in other fields, but it is an example that is important historically and in our opinion pleasant to describe. The species (i.e. rabbits) starting from an initial set of individuals grows exponentially (see Table 6.1). If we do not even consider death, this results in the famous Fibonacci sequence. As $t \to \infty$ we have that the ratio of two consecutive terms approaches a constant value that is $\lim_{t\to\infty} n(t) = \phi n(t-1)$ (such a value of ϕ becomes more and more similar to the "golden ratio", $\phi^G = \frac{1+\sqrt{5}}{2} \simeq 1.61803...$, a number particularly used by architects and artists to mimic the beauty of nature). We can then write

$$\frac{dn}{dt} = \frac{n(t+1) - n(t)}{1} \simeq (\phi - 1)n(t), \tag{6.1}$$

whose solution is $n(t) = n(t_0)e^{(\phi-1)(t-t_0)}$.

Compute the Fibonacci sequence

```
def fibonacci(sequence_length):
    "Return the Fibonacci sequence of length *sequence_length*"
    sequence = [0,1]
    if sequence_length < 1:
        print "Fibonacci sequence only defined for length 1 \
               or greater"
        return
    if 0 < sequence_length < 3:
        return sequence[:sequence_length]
    for i in range(2,sequence_length):
        sequence.append(sequence[i-1]+sequence[i-2])
    return sequence
```

```
fibonacci(12)

#OUTPUT
[0, 1, 1, 2, 3, 5, 8, 13, 21, 34, 55, 89]
```

Two species: Lotka–Volterra equations. The simplest way to describe a more realistic food web is to introduce at least one prey and one predator. Typically in a food web we are overwhelmed by the complexity of the connections, and the field observations necessary even to spot a single predation could take years of field work by several team of ecologists. This situation does not therefore allow, in all the interesting cases, the collection of information on the oscillation in numbers of individuals for every species. For this reason we consider here the case of two species, since it is sufficiently instructive to be considered with care and it can be approached by means of a simple statistical model. Furthermore, this is the simplest example of population dynamics which can be applied in a variety of other situations. The idea is to describe the system by means of a system of coupled differential equations. In the simplest version with one prey y and one predator z, we have

$$\begin{cases} \mathbf{y}'(\mathbf{t}) = \alpha\mathbf{y}(\mathbf{t})\mathbf{z}(\mathbf{t}) - \beta\mathbf{y}(\mathbf{t}) \\ \mathbf{z}'(\mathbf{t}) = \gamma\mathbf{z}(\mathbf{t}) - \delta\mathbf{z}(\mathbf{t})\mathbf{y}(\mathbf{t}). \end{cases} \tag{6.2}$$

For the predator $y(t)$:

- the first term $\alpha y(t)z(t)$ describes the growth based on predation on z. It is an exponential growth (the factor $y(t)$, where the resources are proportional to the prey $\alpha z(t)$;
- the second item $\beta y(t)$ takes into account death or emigration.

For the prey z:

- the first term $\gamma z(t)$ is an exponential growth (assuming infinite and constant resources).
- the second term $\delta z(t)y(t)$ accounts for predation from predators. Since not all the losses by z contribute to the growth of y, this term is different from the first term in the predator equation.

Because of the simplicity of this form and the very general fields of application, many scientists have arrived at similar results (from Pierre-Francois Verhulst in 1838). In this modern formulation they are originated from work by the Italian mathematician Vito Volterra and US statistician Alfred J. Lotka.

An analysis of such a system of equations depends on the values of the parameters $\alpha, \beta, \gamma, \delta$. Generally, though, it is fair to say that the populations of predators and prey oscillate in time, with larger oscillations for prey.

Solve the Lotka–Volterra differential equations

```
from sympy import *

#defining the variables and parameters
var('y z')
var('alfa beta gamma delta', positive=True)
#defining the equations
dy = alfa*z*y - beta*y
dz = gamma*z - delta*z*y
#solving the Lotka Volterra equations
(y0, z0), (y1, z1) = solve([dy, dz], (y, z))
A = Matrix((dy, dz))
print A
#computing the Jacobian
Jacobian = A.jacobian((y, z)); Jacobian
B = Jacobian.subs(y, y0).subs(z, z0)
C = Jacobian.subs(y, y1).subs(z, z1)
print B,C
#stability of the fixed points
solutionB=B.eigenvals()
solutionC=C.eigenvals()
print solutionB,solutionC

#OUTPUT
{gamma: 1, -beta: 1}
{-I*sqrt(beta)*sqrt(gamma): 1, I*sqrt(beta)*sqrt(gamma): 1}
```

Many species, random competition. When adding more and more species to the system, the set of relationships can be effectively described by means of a predation matrix $\mathbf{A}$ whose entries a_{ij} have the following property

$$a_{ij} = \begin{cases} 1 \text{ if } j \text{ predates } i \\ 0 \text{ otherwise } (j \text{ does not predate } i) \end{cases} \tag{6.3}$$

That is A is the matrix of an oriented graph where the out-degree gives the number of predators (the arrows indicate the flow of nutrients). The simplest hypothesis that can be made, is that we do not knowing anything. We then simply draw the connections randomly with uniform probability p. In its undirected version, this simple model is known as *random graph.* and represents the basic benchmark for any network in nature and/or technology.

6.3 Random graphs

Random graphs (Erdős and Rényi, 1959) represent the benchmark for any real network. In a random graph, once given the number of vertices and the probability p to draw an edge between a couple of them, we obtain a graph where connections are assigned randomly. Indeed all the vertices are equal and the probability p does not depend on the vertices involved. In random graphs, then, on average all vertices look the same, and the behaviour of network quantities is well represented by their average quantities. From a mathematical point of view we can start by assigning N vertices, and then check (once for an undirected graph) what is the probability that one vertex i is connected with another vertex j. After $N-1$ checks we can compute the probability that a vertex i (and therefore any vertex in the graph) has degree k. To compute this quantity, let us consider that we must have a link extraction k times (happening with probability p) and no extraction $(N-1-k)$ times (happening with probability $(1-p)$). We do not care about the order in which the k edges are drawn (i.e. the first k times we draw an edge, and the rest of $N-1-k$, we do not), so we must consider all the $\frac{(n-1)!}{(n-1-k)!k!}$ permutations of k successful events and $N-1-k$ unsuccessful events in a row of $N-1$ trials. We then have

$$P_k = \frac{(n-1)!}{(n-1-k)!k!} p^k (1-p)^{n-1-k} = \binom{n-1}{k} p^k (1-p)^{n-1-k}. \tag{6.4}$$

That is, we find a binomial distribution for the various values of degree k; note that the distribution is automatically normalised since

$$\sum_{k=1,n-1} P_k = (p + (1-p))^{n-1} = 1. \tag{6.5}$$

The limit case of $p = 0$ produces an empty graph, while the opposite limit of $p = 1$ produces the complete graph; any intermediate value of p produces a graph with various densities, a situation similar to that of percolation (Stauffer and Aharony, 1994). In the limit $N \to \infty$ and $p \to 0$ with the product pN constant, the above distribution is well represented by a Poisson distribution with mean value $\mu = pN$ and variance $\sigma = pN$. A huge amount of analytical work has been done with random graphs (Bollobás, 1979; Bollobás, 1985) and those analytical results constitute the basic benchmark to evaluate the behaviour of any other graph. Indeed, in random graphs, no correlation between vertices is present, therefore they constitute a sort of "zeroth" hypothesis against which one can measure any cost function relative to real networks. As regards all the other quantities that characterise a network such as distribution of clustering or assortativity (measured as the average degree of the neighbours of a given node), they also display typical behaviour. If the graph is sufficiently sparse the clustering can be computed by checking what the probability is that vertices j and k are connected if they are both neighbours of a vertex i. As a first approximation this is simply given by the probability of drawing a link, i.e. p. So as the size grows ($N \to 0$) and the probability goes to zero ($p \to 0$), so does the clustering. With similar reasoning one can imagine that the average degree of vertex neighbours will have the expected value of $\langle k \rangle = pN$. While the average degree does not capture all the complexity of the

assortativity (Mussmann *et al.*, 2014), there is no reason to consider a random graph to be inhomogeneous.

Generating Random Networks (Erdős–Rényi)

```
import networkx as nx
import random

Number_of_nodes=10
p=0.4

G=nx.Graph()
for n in range(Number_of_nodes):
    G.add_node(n)

node_list=G.nodes()

#generate the graph adding ad edge for each possible couple of nodes
for i1 in range(len(node_list)-1):
    for i2 in range(i1+1,len(node_list)):
        if random.random()<p:
            G.add_edge(node_list[i1],node_list[i2])

pos=nx.circular_layout(G)
nx.draw(G, pos,with_labels=True)
```

Given the experimental evidence that most real networks have instead a different degree distribution, various modifications have been proposed (Chung *et al.*, 2003) to the original model. For example, we can assign a given expected degree sequence $w = (w_1, w_2, ..., w_n)$ for a graph, finding that the probability p_{ij} of having a link between vertices i and j is given by $p_{ij} = \frac{w_i w_j}{\sum_i w_i}$. By construction, the above approach allows us to reproduce a variety of different behaviours, but does not necessarily allow us to reproduce the real complexity of a graph, with correlation between vertices made evident by non-trivial values of assortativity and clustering.

6.3.1 Randomising a graph

Real networks display a long-range level of correlation. Not only is the degree distribution scale-free, but also higher-order correlations are present. It is then useful to have a method of characterising how much these correlations count with respect to a null case. This means that we would like to measure the properties of a real graph against another one where correlations are not present. This can be done by realising a randomisation of the initial graph, not at the level of the degree (a random graph with the same number of vertices and edges of a real network is very likely not to have the same degree sequence), but at a higher level. A procedure to destroy correlation

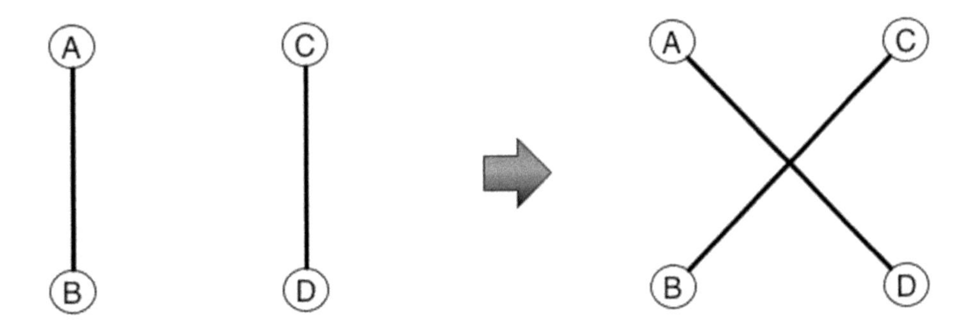

Fig. 6.1 A process of randomisation of a graph, consisting of the swapping of the two end vertices.

by preserving degree sequence can easily be outlined as follows (Maslov *et al.*, 2004). Extract one couple of edges (A, B) and (C, D) with four different vertices (A, B, C, D) and swap the extremes (end vertices) of the two edges (i.e. (A-B), (C-D) $\rightarrow$ (C-B), (A, D)) as outlined in Fig. 6.1 One limitation of this model, seldom pointed out, is that in order to effectively change the set of connections, the network must be sufficiently sparse. If the graph is complete there is obviously nothing to swap(Zlatić *et al.*, 2009). Therefore a randomization is only possible when the number of possible configurations accessible to the algorithm is large enough.

Randomising graphs

```
number_of_swaps=2

while number_of_swaps>0:
    #pick at random a couple of edges and verify
    #they don't share nodes
    edges_to_swap=random.sample(G.edges(),2)
    e0=edges_to_swap[0]
    e1=edges_to_swap[1]

    if len(set([e0[0],e0[1],e1[0],e1[1]]))<4: continue

    #check if the edge already exists and eventually add it
    if not G.has_edge(e0[0],e1[1]):
        G.add_edge(e0[0],e1[1])
    G.remove_edge(e0[0], e0[1])
    if not G.has_edge(e0[1],e1[0]):
        G.add_edge(e0[1],e1[0])
    G.remove_edge(e1[0], e1[1])
```

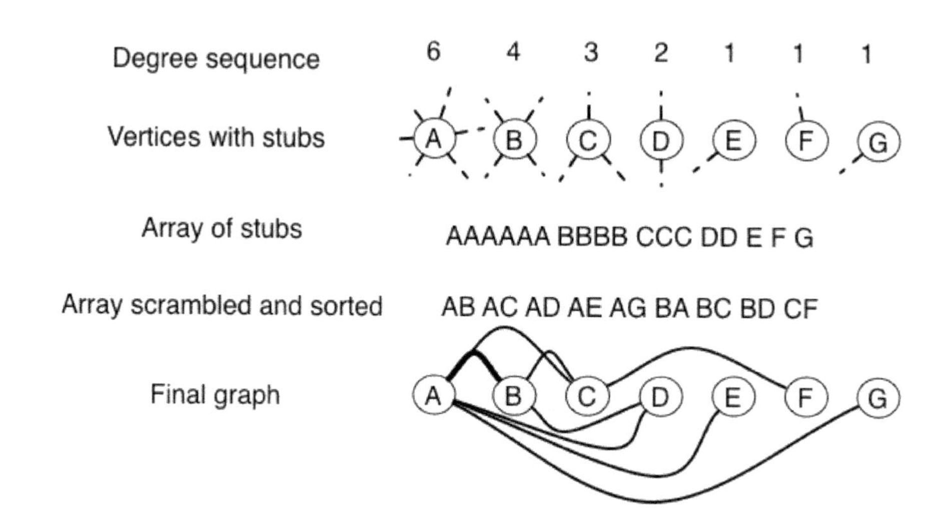

Fig. 6.2 The degree sequence, the vertices with edge stubs, and a possible solution. Note the width of the edge connecting AB, which is counted twice. As previously explained, multi-edges and self-loops are always possible.

```
        number_of_swaps-=1

    pos=nx.circular_layout(G)
    nx.draw(G, pos,with_labels=True)
```

6.4 Configuration models

The importance of random graphs is due to the fact that they represent an instance of being a graph when no particular correlation in the vertices is present. However, correlation in a graph is not only restricted to the degree, that is, to the number of edges per vertex. Actually higher-order correlations are always present, so that phenomena like the "rich club effect" (all the hubs are connected with each other) take place. At this point to have a benchmark of "randomness" given a degree sequence, we need a way to build the "most random" graph with a fixed series of degrees. This is the configuration model (Bender and Canfield, 1978; Molloy and Reed, 1995) which allows us to build a a graph starting from the basic constituents, the vertices with their given edge stubs. Of course not all degree sequences that one can imagine result in a simple graph. For example a degree sequence whose sum is odd, cannot be taken from a simple graph (the sum of all the degrees is always twice the number of edges, i.e. it is an even number). In other cases it must still be possible to build a simple graph, but the solution can be computationally hard to find. An efficient algorithm for building a network with a uniformly drawn random matching through the vertices is sketched here and shown in Fig. 6.2.

- order the vertices from the one with the largest degree to that with the smallest;
- assign to any of them a number of edge stubs given by the degree sequence;
- write down an array whose entries are all the stubs indicated by the label of the vertex to which they belong;
- scramble the array exchanging the position of the entries;
- take the entries in block for two and draw an edge between the vertices indicated.

Configuration model

```
degree_sequence=[6, 4, 3, 2, 1, 1, 1]

#this generate the list of uppercast chars as labels for the nodes
uppercase_char_list=[chr(i) for i in xrange(65,91)]

degree_sequence.sort(reverse=True)
#degree sequence
print "degree sequence:",degree_sequence

stub_list=[]

for deg in degree_sequence:
    label=uppercase_char_list.pop(0)
    for stub in range(deg):
        stub_list.append(label)


print "ordered stub labels",stub_list

random.shuffle(stub_list)

print "shuffled stub labels",stub_list

MG = nx.MultiGraph()

while stub_list!=[]:
    node1=stub_list.pop(0)
    node2=stub_list.pop(0)
    MG.add_edge(node1,node2)

print "graph edge list:",MG.edges()

#OUTPUT
degree sequenece: [6, 4, 3, 2, 1, 1, 1]
ordered stub labels ['A', 'A', 'A', 'A', 'A', 'A', 'B', 'B',
```

```
'B', 'B', 'C','C', 'C', 'D', 'D', 'E', 'F', 'G']
shuffled stub labels ['F', 'B', 'B', 'A', 'C', 'A', 'D', 'A',
'B', 'A', 'A', 'E', 'A', 'C', 'D', 'G', 'C', 'B']
graph edge list: [('A', 'C'), ('A', 'C'), ('A', 'B'), ('A', 'B'),
('A', 'E'), ('A', 'D'), ('C', 'B'), ('B', 'F'), ('D', 'G')]
```

6.5 Gravity model

In the case of the World Trade Web, a simple model to describe the properties of the structure is given by the gravity model. Inspired by the Newtonian law of forces, one can imagine that the flow of trade between two countries might depend on the two "masses" of the countries involved and might be inversely proportional (not necessarily with a square power) to their reciprocal distance. This model, known as "gravity model" (Tinbergen, 1962; Pöyhönen, 1963; Linnemann, 1966) has been widely used with a variety of choices for mass and its form is given by

$$t_{ij} = G(M_i^{\beta_1} M_j^{\beta_2} / D_{ij}^{\beta_3}). \tag{6.6}$$

In this formula, T_{ij} indicates the trade between two countries i, j; the masses $M_{i,j}$ are characteristic of the countries; i, j, $D_{i,j}$ is their reciprocal distance (between the capitals, see for example Fig. 6.3); and G is a constant. There are various explanations for the meaning of the masses $M_{i,j}$ for the countries i, j. The simplest choice is to use the GDP of countries as the quantity M describing the trade F between two nations. This happens since, on the basis of empirical observations that trade is related to abundance of products and that trade between two medium-sized countries should exceed trade between a small and a large country, it has been argued that income should play a crucial role (Helpman and Krugman, 1985).

The values of the exponents $\beta_{1,2,3}$ are determined from data by means of various methods of fitting.

Gravity model

```
import scipy.optimize as optimization
import numpy

# Generate artificial data = straight line with a=0 and b=1
# plus some noise.
xdata = numpy.array([0.0,1.0,2.0,3.0,4.0,5.0])
ydata = numpy.array([0.1,0.9,2.2,2.8,3.9,5.1])

sigma = numpy.array([1.0,1.0,1.0,1.0,1.0,1.0])

# Initial guess.
x0    = numpy.array([0.0, 0.0, 0.0])
```

Fig. 6.3 Distances between countries are computed between the two capitals. Different distances and "masses" give rise to different trades.

```
#defining the gravitational function
def func(x, a, b, c):
    return a + b*x + c*x*x

#optimization
import scipy.optimize as optimization

print optimization.curve_fit(func, xdata, ydata, x0, sigma)

#OUTPUT
(array([ 0.1        ,  0.88142857,  0.02142857]),
       array([[ 0.02753741, -0.0197551 ,  0.0029932 ],
       [-0.0197551 ,  0.02436463, -0.0044898 ],
       [ 0.0029932 , -0.0044898 ,  0.00089796]]))
```

6.6 Fitness model

Along a similar line of reasoning to that for the gravity model, there is a whole class of network models that are based on some "mass" or quantity characteristic of the vertices. This quantity indicated as "fitness" determines the property of the network, starting from the degree of every node. More particularly, the probability of drawing an edge is a function of the fitness of the vertices involved. In more detail, we can assign a real value x_i to every vertex i of the graph. This value can be taken from data (i.e. it could be the income of a person or the daily volume activity of a bank) or extracted from a probability distribution $\rho(x)$. The second step is to determine a function $f(x_i, x_j)$ giving the probability $p_{i,j}$ with which the two vertices (i, j) will be connected (Caldarelli *et al.*, 2002; Boguñá and Pastor-Satorras, 2003; Servedio *et al.*, 2004). According to the various possible choices for both the distribution $\rho(x)$ and the linking probability $p_{i,j}$ different classes of networks appear.

Fitness model

```
import math

G=nx.Graph()

#this is our z(N)
ave_value=1.0
N=5000

def fitness_function():
    return random.expovariate(4.0/ave_value)

def generate_function(x1,x2):
    if x1+x2-ave_value<0.0:
        return 0
    else:
        return 1

for n in range(N):
    G.add_node(n,fitness=fitness_function())

node_list=G.nodes()

#generate the graph adding ad edge for each possible couple of nodes
for i1 in range(len(node_list)-1):
    for i2 in range(i1+1,len(node_list)):
        x1=G.node[node_list[i1]]['fitness']
        x2=G.node[node_list[i2]]['fitness']
        if generate_function(x1,x2)==1:
```

```
        G.add_edge(node_list[i1],node_list[i2])

degree_sequence=sorted(nx.degree(G).values(),reverse=True)

hist(degree_sequence,bins=15)
```

To fix our ideas consider, for example, the expected value $k(x)$ of the degree of a vertex whose fitness is x, we have

$$k(x) = N \int_0^\infty \rho(y) f(x,y) = NF(x). \tag{6.7}$$

If the function is a monotonous one, we can invert it and in the limit of large N (where we can neglect finite corrections) we can write

$$P(k) = \rho \left[F^{-1}(\frac{k}{n}) \right] \frac{d}{dk} F^{-1}(\frac{k}{n}). \tag{6.8}$$

If the fitnesses are exponentially distributed so that x_i, x_j are extracted from a $\rho(x) = Ae^{-x}$, where $A = 1$ is the normalisation constant and the linking probability $p_{i,j}$ is a threshold function $p_{i,j} = \theta(x_i + x_j - z(N))$, we can obtain a scale-free network $P(k) \propto k^{-2}$. Of course, since from (6.8) we see that the form of the $P(k)$ is that of the $\rho(x)$, for all cases in which F can be invertible, we find that whenever the fitnesses are power-law distributed (Pareto's law) the network will have a scale-free distribution.

Various successful applications of this model have been made for economic and financial networks such as the World Trade Web (Garlaschelli and Loffredo, 2004*a*).

6.7 Barabási–Albert model

While the number of countries and their reciprocal distance can be considered to be (at least for the latter) fairly stationary in time, this is not the case for routers or web sites. In such cases a better suited statistical model is necessary. The most successful model of graph growth based on this totally different approach is the Barabási–Albert (BA) model (Barabási and Albert, 1999; Albert and Barabási, 2002). In this model the number of vertices varies continuously in time; in particular, time is discretised and at every time step a new vertex is added. A simple procedure for building a BA network is to start from a small initial graph and for every time step:

- introduce a new vertex in the system
- link the vertex to the initial graph by drawing m new edges; the destination vertices are chosen with a probability proportional to their degree. This means that

$$p = p(k_i) = \frac{k_i}{\sum_{j=1,n} k_j}. \tag{6.9}$$

Note that, since at every time step only one vertex enters, we have for the order n, that is, the number of vertices and the size m of the network, respectively

$$n = n_0 + t,$$
$$m = \frac{1}{2} \sum_{j=1,n} k_i = mt + m_0. \tag{6.10}$$

Iteration of this simple procedure gives rise to scale-free networks characterised by a power law in the degree distribution of the kind $P(k) \propto k^{-3}$. We can provide some scaling arguments to derive such behaviour. The first step is in deriving the law of growth for the degree of a given vertex.

As a first approximation let us consider the degree as a continuous variable. The degree of any vertex can only increase, the variation of the degree of one node in one time step will be given by how many edges we add (m) times the probability $p(k)$ to get an edge from the newcomer vertex just added. This means

$$\frac{\partial k_i}{\partial t} = m_0 \Pi(k) = m \frac{k_i}{\sum_{j=1,n} k_j} = \frac{m k_i}{2mt + m_0}. \tag{6.11}$$

If we assume that the initial edges are 0 (or in any case we study the behaviour for large t), we have

$$\frac{\partial k_i}{\partial t} = \frac{k_i}{2t} \rightarrow k_i(t) = m \left(\frac{t}{t_i} \right)^{1/2}, \tag{6.12}$$

where t_i is the time at which the vertex entered the system. The younger it is, the smaller the degree.

From this first result (the degree grows with the square root power of time) we can derive the form of the degree distribution. The basic idea is that we can use time and degree interchangeably, this means that the probability $P(k_i < k)$ that a vertex has a degree lower than k is $P(k_i < k) = P(t_i > \frac{m_0^2 t}{k^2})$. We also know that vertices enter at a constant rate so that the time distribution is uniform in time, that is $P(t) = A$, where the constant value of A can be determined by imposing $A = \int_0^n P(t) = An = 1$, so that $A = 1/n = 1/(n_0 + t)$. In this way, we write

$$P(t_i > \frac{m^2 t}{k^2}) = 1 - P(t_i \leq \frac{m^2 t}{k^2}) = 1 - \frac{m^2 t}{k^2} \frac{1}{(n_0 + t)}, \tag{6.13}$$

from which we obtain

$$P(k) = \frac{\partial P(k_i > k)}{\partial k} = \frac{2m^2 t}{(n_0 + t)} \frac{1}{k^3}. \tag{6.14}$$

Therefore, we find that the degree distribution is a power law with a value of the exponent $\gamma = 3$.

Barabási–Albert model

```
N0=6
p=0.6
new_nodes=1000

G=nx.gnp_random_graph(N0, p)

for eti in range(new_nodes):
    m=3
    new_eti="_"+str(eti)
    target_nodes=[]
    while m!=0:
        part_sum=0.0
        rn=random.random()
        for n in G.nodes():
            base=part_sum
            step=part_sum+G.degree(n)/(G.number_of_edges()*2.0)
            part_sum=part_sum+G.degree(n)/(G.number_of_edges()*2.0)
            if rn>=base and rn<step:
                if n in target_nodes: break
                target_nodes.append(n)
                m=m-1
                break

    for n_tar in target_nodes:
        G.add_edge(new_eti,n_tar)

degree_sequence=sorted(nx.degree(G).values(),reverse=True)

hist(degree_sequence,bins=15)
```

6.8 Reconstruction of networks

Finally, in this part devoted to computer simulations, we present a way to reconstruct graphs. Reconstructing a network when only limited information is available represents one of the major challenges in the field of complex systems, especially those concerning socio-economics: prominent examples are provided by financial systems (for which the information that regulators are able to collect is limited by confidentiality issues), social systems (for which the amount of available information is strongly limited either by the tutelage of privacy or by the unfeasibility of exhaustive sampling), and ecosystems (for which collecting data about all the relationships between species is often unfeasible). The importance of knowing the structure of such systems becomes evident upon considering that, e.g., a correct estimation of the resilience of interbank networks to financial shocks rests heavily upon knowledge of the detailed inter-dependencies among such institutions. Similarly, a detailed understanding of disease spreading across a population is based on knowledge of the structure of the underlying social network.

Measuring likelihood of unknown configurations

When dealing with network reconstruction, two issues are crucial: (1) making optimal use of the available information, and (2) drawing as unbiased as possible conclusions on the unknown portion of the system. Both issues can be dealt with by adopting the so-called MaxEnt (i.e. maximum entropy) principle, which results in a constrained maximisation of Shannon entropy (Shannon, 1948; Jaynes, 1957),

$$S = -\sum_G P(G) \ln P(G), \tag{6.15}$$

where the constraints are represented by expected values of quantities of interest for the problem at hand, $C_a^* = \sum_G P(G) C_a(G)$, beside the normalization condition $1 = \sum_G P(G)$. Such a procedure allows one to obtain the probability $P(G)$ that a particular network configuration G is realized (1) upon preserving the available information on G, (2) while randomising everything else. Maximising S with the constraints imposed by the available information leads to a probability distribution, $P(G)$, defined over the space of possible graphs, i.e.

$$P(G) = \frac{e^{-H(G)}}{Z} = \frac{e^{-\sum_a \theta_a C_a}}{Z}, \tag{6.16}$$

where $H(G)$ sums up the available information on the network G and $Z = \sum_G e^{-H(G)}$ normalises the probability coefficient to one. Such a theoretical framework (also known as "exponential random graph" framework (Park and Newman, 2004)) can be used to determine the probability of observing any kind of network configuration (binary or weighted, undirected or directed, monopartite or bipartite, monoplex or multiplex) from the available information we have about it. This means that such a framework can be used to define statistical benchmarks (or, equivalently, null models) with which to compare our observations, in order to decide how significant (i.e. informative) they are. In practice, the observed value of a generic quantity $X(G)$ can be compared with its ensemble expectation $\langle X \rangle = \sum_G P(G) X(G)$, and the discrepancy between the two can be estimated via the standard deviation $\sigma[X] = \sqrt{\langle X^2 \rangle - \langle X \rangle^2}$ (or, more compactly, through the z-score $z_X = (X(G) - \langle X \rangle)/\sigma[X]$).

Noteworthy cases are represented by probability distributions which factorise into products of link-specific probabilities, p_{ij}. The most famous example is represented by the configuration model, in which the available information coincides with the knowledge of the number of connections (i.e. the degree k_i^*, $\forall i$) for every node in the observed graph G^*. Such information, in the context of the aforementioned MaxEnt principle, determines the ensemble probability distribution of the network G in (6.16), such that $H(G) = \sum_i \theta_i k_i(G)$, where the parameters θ_i, $\forall\, i$ play the role of Lagrange multipliers. In this case, $P(G)$ factorises into the product $P(G) = \prod_{i<j} p_{ij}^{g_{ij}} (1 - p_{ij})^{1-g_{ij}}$ with the adjacency matrix element $g_{ij} = 0, 1$ and $p_{ij} = x_i x_j / (1 + x_i x_j)$ (having defined

$x_i = e^{-\theta_i}$). The Chung-Lu model (Chung and Lu, 2002), is a simplified version of the configuration model, valid for sparse networks (i.e. whenever $x_i x_j \ll 1$, $p_{ij} \simeq x_i x_j$).

6.8.1 Estimating the unknown parameters

The MaxEnt principle leaves us with the problem of estimating the unknown parameters $\theta_i, \forall i$. A typical approach prescribes maximising the probability of observing precisely the configuration G^* under analysis (Garlaschelli and Loffredo, 2008; Squartini and Garlaschelli, 2011). This translates into maximisation of the likelihood function $\mathcal{L}(\vec{\theta}) = \ln P(G^*)$ (which can be considered as a function of the unknown parameters $\theta_i, \forall i$), which implies solving the system of equations

$$\frac{\partial \mathcal{L}(\vec{\theta})}{\partial \theta_i} = 0, \; \forall\, i. \tag{6.17}$$

The algorithm for estimating the parameters of the configuration model reads $k_i^* = \sum_{j(\neq i)} x_i x_j / (1 + x_i x_j), \forall\, i$. Notice that, in this case, the probability that any two nodes i and j establish a connection is driven by the value of their degrees. Other relevant examples are provided by the weighted configuration model (Squartini and Garlaschelli, 2011), which prescribes that the connection probability between any two nodes is driven by their strengths, and the Reciprocal Configuration Model (Squartini and Garlaschelli, 2011), which accounts for the number of reciprocal interactions each node establishes.

6.8.2 From null models to reconstruction methods

What emerges from our discussion is that by raising the amount of structural information used to determine $P(G)$, progressively more accurate estimates of the configuration of a particular network G^* are obtained. The baseline Erdős Renyi random graph is recovered as the simplest network model, in which the only available information on the network is the total number of links, $L(G^*)$. Indeed, the Erdős–Renyi model is so simple that it can hardly be imagined to be useful in reconstructing (i.e. reproducing) a real network.

The amount of structural information used to determine $P(G)$ must thus be increased. As a consequence, many algorithms have privileged the use of weighted information (such as the node-specific total weights, i.e. the strength sequence), resting upon the assumption that it would have been more informative than the purely topological information provided by, e.g., the number of neighbours of each node. However, as shown in (Squartini *et al.*, 2011*a*; Squartini *et al.*, 2011*b*), constraining the strength sequence alone induces a network which is much denser than that observed, thus completely failing in reproducing its skeleton.

The most straightforward way to restore such missing information is by combining some information on the topology of the network with some information on its weighted structure. Operationally, this translates into simultaneously constraining the node degrees $k_i^*, \forall\, i$ and the node strengths $s_i^*, \forall\, i$; such a prescription defines the so-called Enhanced Configuration Model (ECM) (Mastrandrea *et al.*, 2014*a*). More explicitly, the ECM formulas resulting from the MaxEnt principle for estimating the probability that nodes i and j are connected, and the corresponding expected weight are

$$p_{ij} = \frac{x_i x_j y_i y_j}{1 + x_i x_j y_i y_j - y_i y_j}, \ \langle w_{ij} \rangle = \frac{p_{ij}}{1 - y_i y_j}, \tag{6.18}$$

whose unknown parameters can be estimated through solving the system

$$\begin{cases} k_i = \sum_{j(\neq i)} p_{ij}, \forall\, i \\ s_i = \sum_{j(\neq i)} \langle w_{ij} \rangle, \forall\, i. \end{cases} \tag{6.19}$$

Upon interpreting a link-specific weight as a sum of single links, the ECM clarifies the structural role played by the first link established between any two nodes with respect to those following, pointing out their asymmetric role. In other words, the ECM tells us that *creating* a connection and *reinforcing* an existing connection are processes obeying different rules, driving nodes towards integration or segregation. It can be shown that the magnitude of the product $x_i x_j$ discriminates between these two tendencies, further highlighting the prominent role played by purely topological quantities (Mastrandrea *et al.*, 2014*a*). The explanatory power of the ECM has been proved upon analysing the World Trade Web, whose entire multiplex structure can be reproduced by constraining the layer-specific degree and strength sequences of the world countries (Mastrandrea *et al.*, 2014*b*).

6.8.3 The fitness model

As previously said, the unknown parameters defining $P(G)$ can be estimated through the maximisation of the likelihood function $\mathcal{L}(\vec{\theta})$. However, networks exist for which the probability that any two nodes interact can be explicitly written in terms of non-structural quantities which are typical of the system under analysis. A prominent example is provided by the World Trade Web: the gross domestic product (GDP) of countries has been shown to be the node-specific characteristic which shapes their structure and drives their evolution (Garlaschelli and Loffredo, 2004*a*; Cimini *et al.*, 2015*c*). In cases like these, the unknown parameters can be substituted by known, intrinsic, node-specific fitnesses. For the World Trade Web we have

$$p_{ij}^{WTW} = \frac{z\, GDP_i\, GDP_j}{1 + z\, GDP_i\, GDP_j}, \tag{6.20}$$

where the extra parameter z can be tuned to reproduce the observed number of links, $L(G) = \sum_i \sum_{j(\neq i)} p_{ij}^{WTW}$. In this case, the fitness model can be interpreted as an approximation of the Configuration Model, for which a relation of the kind $x_i = \sqrt{z} GDP_i$ has been verified (Garlaschelli and Loffredo, 2004*a*; Cimini *et al.*, 2015*c*; Cimini *et al.*, 2015*a*; Cimini *et al.*, 2015*b*). Although the WTW-specific fitness model reproduces very accurately the topological structure of such a network (Garlaschelli and Loffredo, 2004*a*), it does not provide any information on its weighted structure which must be estimated with a different method (e.g. see (Almog *et al.*, 2015)).

6.8.4 Bootstrapping a network.

The steps discussed so far work when, e.g., the degree or the strength of each node is known. It is noteworthy, this approach can be extended to deal with cases where the aforementioned information is known for only some of the nodes. As an example, let us imagine that the degree sequence of only a subset of nodes, I, is known: in this case, we cannot solve the system of equations coming from maximising the likelihood. Thus, the node-specific missing information must be replaced by some known fitness, g_i, $\forall\, i$, obtaining connection probabilities formally analogous to those coming from the fitness model: $\tilde{p}_{ij} = zg_ig_j/(1 + zg_ig_j)$. At this point, a "reduced" likelihood estimation can be carried out, in order to determine z (Cimini *et al.*, 2015*c*; Musmeci *et al.*, 2013):

$$\sum_{i \in I} k_i = \sum_{i \in I} \sum_{j(\neq i)} \tilde{p}_{ij}. \tag{6.21}$$

Although such an estimation is purely based on the known portion of topological information, i.e. the degree sequence of nodes in I, the obtained value of z determines the connection probabilities of all pairs of nodes. Notably, (6.21) works even when the degree of only one node is known. As for the case where complete knowledge is available, $P(G)$ can be estimated as $P(G) = \prod_{i<j} \tilde{p}_{ij}^{g_{ij}} (1-\tilde{p}_{ij})^{1-g_{ij}}$; the observed value of a generic quantity $X(G)$ can thus be compared with its expectation $\langle X \rangle$ defined by the measure $\tilde{p}_{ij}$, and the error can be estimated through the corresponding $\sigma[X]$.

When applied to the WTW and to the E-mid interbank network, the boostrap method performs remarkably well, even when considering a subset of only a few nodes, for estimating the parameter z (Musmeci *et al.*, 2013). Clearly, the accuracy of the estimates increases with the size of the subset about which we have information; however, it has been shown that knowing 10% of the nodes is enough in order to obtain remarkably good agreement, for both the WTW and the E-mid interbank network (Cimini *et al.*, 2015*c*; Musmeci *et al.*, 2013).

References

Adamic, Lada A. and Glance, Natalie (2005). The political blogosphere and the 2004 U.S. election. In *Proceedings of the 3rd international workshop on Link discovery - LinkKDD '05*, New York, New York, USA, pp. 36–43. ACM Press.

Albert, Réka and Barabási, Albert-László (2002). Statistical mechanics of complex networks. *Reviews of Modern Physics*, **74**(1), 47–97.

Albrecht, S., Lubcke, M., and Hartig-Perschke, R. (2007). Weblog campaigning in the German Bundestag election 2005. *Social Science Computer Review*, **25**(4), 504–520.

Almog, Assaf, Squartini, Tiziano, and Garlaschelli, Diego (2015). A GDP-driven model for the binary and weighted structure of the International Trade Network. *New Journal of Physics*, **17**(1), 013009.

Bachelier, Louis (1900). Théorie de la spéculation. *Annales Scientifiques de lÉcole Normale Supérieure*, **3**(17), 21–86.

Balassa, Béla (1965). Trade liberalization and revealed comparative advantage. *Manchester School*, **33**, 99–123.

Barabási, Albert-László and Albert, Reka (1999). Emergence of scaling in random networks. *Science*, **286**(5439), 509–512.

Barabási, Albert-László and Bonabeau, Eric (2003). Scale-Free Networks. *Scientific American*, **288**, 60–69.

Baran, Paul (1960). Reliable Digital Communications Systems Using Unreliable Network Repeater Nodes. Technical report, Rand Corporation.

Baran, Paul (1964). On Distributed Communications. Technical report, Rand Corporation.

Barigozzi, Matteo, Fagiolo, Giorgio, and Garlaschelli, Diego (2010). Multinetwork of international trade: A commodity-specific analysis. *Physical Review E*, **81**(4), 046104.

Barrat, Alain, Barthélemy, Marc, Pastor-Satorras, Romualdo, and Vespignani, Alessandro (2004). The architecture of complex weighted networks. *Proceedings of the National Academy of Sciences of the United States of America*, **101**(11), 3747–3752.

Barrat, Alain, Barthélemy, Marc, and Vespignani, Alessandro (2008). *Dynamical Processes on Complex Networks*. Cambridge University Press, Cambridge (UK).

Barthélemy, Marc (2004). Betweenness centrality in large complex networks. *European Physical Journal B - Condensed Matter*, **38**(2), 163–168.

Bender, Edward A. and Canfield, E. Rodney (1978). The asymptotic number of labeled graphs with given degree sequences. *Journal of Combinatorial Theory, Series A*, **24**(3), 296–307.

Berners Lee, Tim (2007). Giant Global Graph. Technical report.

Boccaletti, Stefano, Latora, Vito, Moreno, Yamir, Chavez, Martin, and Hwang, D.-

U. (2006). Complex networks: structure and dynamics. *Physics Reports*, **424**(45), 175–308.

Boguñá, Marian and Pastor-Satorras, Romualdo (2003). Class of correlated random networks with hidden variables. *Physical Review E*, **68**, 036112.

Boldi, Paolo and Vigna, Sebastiano (2014). Axioms for centrality. *Internet Mathematics*, **10**, 222–262.

Bollobás, Béla (1979). *Graph Theory, An Introductory course* (1 edn). Springer, New York.

Bollobás, Béla (1985). *Random Graphs.* Academic Press, London.

Brandes, Ulrik (2001). A faster algorithm for betweenness centrality. *Journal of Mathematical Sociology*, **25**, 163–177.

Broder, Andrei, Kumar, Ravi, Maghoul, Farzin, Raghavan, Prabhakar, Rajagopalan, Sridhar, Stata, Raymie, Tomkins, Andrew, and Wiener, Janet (2000). Graph structure in the web. *Computer Networks*, **33**, 309–320.

Caldarelli, Guido (2007). *Scale-free networks: complex webs in nature and technology.* Oxford University Press, Oxford.

Caldarelli, Guido, Battiston, Stefano, Garlaschelli, Diego, and Catanzaro, Michele (2004). Emergence of complexity in financial networks. *Lecture Notes in Physics*, **650**, 399–423.

Caldarelli, Guido, Capocci, Andrea, De Los Rios, Paolo, and Muñoz, Miguel-Angel (2002). Scale-free networks from varying vertex intrinsic fitness. *Physical Review Letters*, **89**(25), 258702.

Caldarelli, Guido, Chessa, Alessandro, Pammolli, Fabio, Pompa, Gabriele, Puliga, Michelangelo, Riccaboni, Massimo, and Riotta, Gianni (2014). A multi-level geographical study of Italian political elections from Twitter data. *PloS one*, **9**, e95809.

Caldarelli, Guido, Cristelli, Matthieu, Gabrielli, Andrea, Pietronero, Luciano, Scala, Antonio, and Tacchella, Andrea (2012). A network analysis of countries' export flows: firm grounds for the building blocks of the economy. *PloS one*, **7**(10), e47278.

Caldarelli, Guido, Higgs, Paul, and Mckane, Alan (1998). Modelling coevolution in multispecies communities. *Journal of theoretical biology*, **193**(2), 345–358.

Callaway, Duncan S., Newman, Mark E. J., Strogatz, Steven H., and Watts, Duncan J. (2000). Network Robustness and Fragility: Percolation on Random Graphs. *Physical Review Letters*, **85**(25), 5468–5471.

Capocci, Andrea, Rao, Francesco, and Caldarelli, Guido (2008). Taxonomy and clustering in collaborative systems: The case of the on-line encyclopedia Wikipedia. *EPL (Europhysics Letters)*, **81**(2), 28006.

Capocci, Andrea, Servedio, Vito D.P., Colaiori, Francesca, Buriol, Luciana S., Donato, Debora, Leonardi, Stefano, and Caldarelli, Guido (2006). Preferential attachment in the growth of social networks: the internet encyclopedia Wikipedia. *Physical Review E*, **74**(3), 36111–36116.

Catanzaro, Michele and Caldarelli, Guido (2012). *Networks A very Short Introduction.* Oxford University Press, Oxford.

Christian, Robert R. and Luczkovich, Joseph J. (1999). Organizing and understanding a winters seagrass foodweb network through effective trophic levels. *Ecological Modelling*, **117**, 99–124.

Chung, Fan and Lu, Linyuan (2002). The average distances in random graphs with given expected degrees. *Proceedings of the National Academy of Sciences USA*, **99**, 15879–15882.

Chung, Fan, Lu, Linyuan, and Vu, Van (2003). Spectra of random graphs with given expected degrees. *Proceedings of the National Academy of Sciences of the United States of America*, **100**(11), 6313–8.

Cimini, Giulio, Squartini, Tiziano, Gabrielli, Andrea, and Garlaschelli, Diego (2015*a*). Estimating topological properties of weighted networks from limited information. *Physical Review E*, **92**(4), 040802.

Cimini, Giulio, Squartini, Tiziano, Garlaschelli, Diego, and Gabrielli, Andrea (2015*b*). Systemic risk analysis on reconstructed economic and financial networks. *Scientific Reports*, **5**, 15758.

Cimini, Giulio, Squartini, Tiziano, Musmeci, Nicolò, Puliga, Michelangelo, Gabrielli, Andrea, Garlaschelli, Diego, Battiston, Stefano, and Caldarelli, Guido (2015*c*). Reconstructing topological properties of complex networks using the fitness model. *Lecture Notes in Computer Science*, **8852**, 323–333.

Cohen, Joel E. (1977). Ratio of prey to predators in community food webs. *Nature*, **270**(5633), 165–167.

Del Vicario, Michela, Bessi, Alessandro, Zollo, Fabiana, Petroni, F., Scala, Antonio, Caldarelli, G., Stanley, H.E., and Quattrociocchi, W. (2016). The spreading of misinformation online. *Proceedings of the National Academy of Science*, **113**, 554–559.

DiGrazia, Joseph, McKelvey, Karissa, Bollen, Johan, and Rojas, Fabio (2013). More tweets, more votes: social media as a quantitative indicator of political behavior. *SSRN Electronic Journal*, **8**, e79449.

Dorogovtsev, Sergey N. (2010). *Lectures on Complex Networks*. Oxford University Press, Oxford.

Dorogovtsev, Sergey N. and Mendes, José Fernando F. (2003). *Evolution of Networks: From Biological Nets to the Internet and WWW*. Oxford University Press, Oxford (UK).

Dunne, Jennifer A., Williams, Richard J., and Martinez, Neo D. (2002). Network structure and biodiversity loss in food webs: robustness increases with connectance. *Ecology Letters*, **5**(4), 558–567.

Eom, Young-Ho, Puliga, Michelangelo, Smailović, Jasmina, Mozetič, Igor, and Caldarelli, Guido (2015). Twitter-based analysis of the dynamics of collective attention to political parties. *PloS one*, **10**, e0131184.

Erdős, Paul and Rényi, Alfred (1959). On random graphs. *Publicationes Mathematicae Debrecen*, **6**, 290–297.

Estrada, Ernesto, Fox, Maria, Higham, Des, and Oppo, Gian Luca (ed.) (2010). *Network Science - Complexity in Nature and Technology*. Springer International Publishing, New York.

Fagiolo, Giorgio, Reyes, Javier A., and Schiavo, Stefano (2009). World-trade web: topological properties, dynamics, and evolution. *Physical Review E*, **79**, 036115.

Frobenius, Georg (1912). Ueber Matrizen aus nicht negativen Elementen. *Sitzungsber. Königl. Preuss. Akad. Wiss*, 456–477.

Garlaschelli, Diego, Di Matteo, Tiziana, Aste, Tomaso, Caldarelli, Guido, and Loffredo, Maria I. (2007). Interplay between topology and dynamics in the world trade web. *The European Physical Journal B*, **57**(2), 159–164.

Garlaschelli, Diego and Loffredo, Maria I. (2004*a*). Fitness-dependent topological properties of the world trade web. *Physical Review Letters*, **93**(18), 188701.

Garlaschelli, Diego and Loffredo, Maria I. (2004*b*). Patterns of link reciprocity in directed networks. *Physical Review Letters*, **93**(26), 268701.

Garlaschelli, Diego and Loffredo, Maria I. (2008). Maximum likelihood: Extracting unbiased information from complex networks. *Physical Review E*, **78**(1), 015101.

Gaulier, Guillaime and Zignago, Soledad (2010). BACI: International Trade Database at the Product-level. Technical report, Centre dEtudes Prospectives et dInformations Internationales.

Girvan, Michelle and Newman, Mark E.J. (2002). Community structure in social and biological networks. *Proceedings of the National Academy of Sciences of the United States of America*, **99**(12), 7821–7826.

Gleditsch, Kristian S. (2002). Expanded Trade and GDP data. *Journal of Conflict Resolution*, **46**, 712.

Goh, K.-I., Kahng, B., and Kim, D. (2001). Universal behavior of load distribution in scale-free networks. *Physical Review Letters*, **87**(27), 278701.

Goldwasser, Lloyd and Roughgarden, Jonathan (1993). Construction and Analysis of a Large Caribbean Food Web. *Ecology*, **74**(4), 1216–1233.

Gonçalves, Bruno, Perra, Nicola, and Vespignani, Alessandro (2011). Modeling users' activity on twitter: validation of dunbar's number. *PLoS One*, **6**, e22656.

Gonzaga, Flavio B., Barbosa, Valmir C., and Xexéo, Geraldo B. (2014). The network structure of mathematical knowledge according to the Wikipedia, MathWorld, and DLMF online libraries. *Network Science*, **2**, 367–386.

Gower, John C. (1966). Some distance properties of latent root and vector methods used in multivariate analysis. *Biometrika*, **53**, 325–338.

Greenwald, Bruce, C.N. and Stiglitz, Joseph E. (1993). Financial market imperfections and business cycles. *Quarterly Journal of Economics*, **108**, 77–114.

Guimerà, Roger, Sales-Pardo, Marta, and Amaral, Luís A Nunes LAN (2004). Modularity from fluctuations in random graphs and complex networks. *Physical Review E*, **70**(2), 025101.

Hall, Stephen J. and Raffaelli, David (1991). Food-web patterns: lessons from a species-rich web. *Journal of Animal Ecology*, **60**(3), 823–841.

Havens, Karl (1992). Scale and structure in natural food webs. *Science*, **257**(5073), 1107–1109.

Helpman, Elhanan and Krugman, Paul (1985). *Market Structure and Foreign Trade.* MIT Press, Cambridge, MA.

Hidalgo, César A. and Hausmann, Ricardo (2009). The building blocks of economic complexity. *Proceedings of the National Academy of Sciences of the United States of America*, **106**(26), 10570–5.

Hidalgo, César A., Klinger, Bertram, Barabási, Albert-László, and Hausmann, Ricardo (2007). The product space conditions the development of nations. *Science*, **317**, 482–487.

Huxham, Mark, Beaney, S., and Raffaelli, David (1996). Do parasites reduce the chances of triangulation in a real food web? *Oikos*, **76**, 284–300.

Jaynes, E. T. (1957). Information Theory and Statistical Mechanics. *Physical Review*, **106**(4), 620–630.

Kleinberg, Jon (1999). Hubs, authorities, and communities. *ACM Computing Surveys*, **31**(4es).

Langville, Amy N. and Meyer, Carl D. (2003). *Google's PageRank and Beyond: The Science of Search Engine Rankings*. Princeton University Press, Princeton, NJ (USA).

Lazer, David, Pentland, Alex, Adamic, Lada, Aral, Sinan, Barabasi, Albert-Laszlo, Brewer, Devon, Christakis, Nicholas, Contractor, Noshir, Fowler, James, Gutmann, Myron, Jebara, Tony, King, Gary, Macy, Michael, Roy, Deb, and Alstyne, Marshall Van (2009, February). Computational Social Science. *Science*, **323**, 721.

Linnaeus, Carl (1735). *Systema Naturae*. Biodiversity Heritage Library, Leiden.

Linnemann, Hans (1966). *An Econometric Study of International Trade Flows*. North-Holland, Amsterdam.

Mandelbrot, Benoit (1963). The variation of certain speculative prices. *Journal of Business*, **36**, 394–419.

Mantegna, Rosario Nunzio (1999). Hierarchical structure in financial markets. *European Physical Journal B*, **11**, 193–197.

Mardia, Kanti V., Kent, John T., and Bibby, John M. (1979). *Multivariate Analysis*. Academic Press, San Diego.

Markowitz, Harry (1952). Portfolio selection. *Journal of Finance* (07), 77–91.

Martin, Owen S. (2011). A wikipedia literature review. *CoRR*, **abs/1110.5863**.

Martinez, Neo D. (1991). Artifacts or attributes? Effects of resolution on the Little Rock Lake food web. *Ecological Monographs*, **61**(4), 367–392.

Martinez, Neo D., Hawkins, B.A., Dawah, H.A., and Feifarek, B.P. (1999). Effects of sampling effort on the characterization of food web structure. *Ecology*, **80**, 1044–1055.

Maslov, Sergei, Sneppen, Kim, and Zaliznyak, Alexei (2004). Detection of topological patterns in complex networks: Correlation profile of the internet. *Physica A*, **333**(1-4), 529–540.

Mastrandrea, Rossana, Squartini, Tiziano, Fagiolo, Giorgio, and Garlaschelli, Diego (2014*a*). Enhanced reconstruction of weighted networks from strengths and degrees. *New Journal of Physics*, **16**, 043022.

Mastrandrea, Rossana, Squartini, Tiziano, Fagiolo, Giorgio, and Garlaschelli, Diego (2014*b*). Reconstructing the world trade multiplex: The role of intensive and extensive biases. *Physical Review E*, **90**(6), 1–18.

Memmott, Jane, Martinez, Neo D., and Cohen, Joel E. (2000). Predators, parasitoids and pathogens: species richness, trophic generality and body sizes in a natural web. *Journal of Animal Ecology*, **69**, 1–15.

Molloy, Mike and Reed, Bruce (1995). A critical point for random graphs with a given degree sequence. *Random Structure and Algorithms*, **6**, 161–180.

Montoya, José M. and Solé, Ricard V. (2002). Small-world patterns in food webs. *Journal of Theoretical Biology*, **214**(3), 405–412.

Muchnik, Lev, Itzhack, Royi, Solomon, Sorin, and Louzoun, Yoram (2007). Self-emergence of knowledge trees: extraction of the Wikipedia hierarchies. *Physical Review E*, **76**(1), 016106.

Musmeci, Nicolò, Battiston, Stefano, Caldarelli, Guido, Puliga, Michelangelo, and Gabrielli, Andrea (2013). Bootstrapping topological properties and systemic risk of complex networks using the fitness model. *Journal of Statistical Physics*, **151**(3-4), 720–734.

Mussmann, Stephen, Moore, John, Pfeiffer, Joseph J., and Neville, Jennifer (2014). Assortativity in Chung Lu Random Graph Models. In *Proceedings of the 8th Workshop on Social Network Mining and Analysis - SNAKDD'14*, New York, New York, USA, pp. 1–8. ACM Press.

Newman, Mark E.J. (2003). The structure and function of complex networks. *SIAM review*, **15**(3), 247–262.

Newman, Mark E.J. (2006). Modularity and community structure in networks. *Proceedings of the National Academy of Sciences of the United States of America*, **103**(23), 8577–8582.

Newman, Mark E.J. (2010). *Networks an Introduction.* Oxford University Press, Oxford.

Newman, Mark E.J. and Girvan, Michelle (2004). Finding and evaluating community structure in networks. *Physical Review E*, **69**(2), 026113.

Page, Lawrence, Brin, Sergey, Motwami, R., Winograd, Terry, and Motwani, Rajeev (1999). The PageRank citation ranking: bringing order to the web.

Park, Juyong and Newman, Mark E.J. (2004). Statistical mechanics of networks. *Physical Review E*, **70**(6), 066117.

Perra, Nicola and Fortunato, Santo (2008). Spectral centrality measures in complex networks. *Physical Review E*, **78**(3), 036107.

Perra, Nicola, Zlatić, Vinko, Chessa, Alessandro, Conti, Claudio, Donato, Debora, and Caldarelli, Guido (2009). Pagerank equation and localization in the www. *EPL (Europhysics Letters)*, **88**, 48002.

Perron, Oskar (1907). Zur Theorie der Matrices. *Mathematische Annalen*, **64**(2), 248–263.

Pimm, Stuart L., Lawton, John H., and Cohen, Joel E. (1991). Food web patterns and their consequences. *Nature*, **350**(6320), 669–674.

Pöyhönen, Pentti (1963). Toward a general theory of international trade. *Ekonomiska. Samfundets Tidskrift, Tredje serien, Argang*, **16**, 69–77.

Serrano, Maria Ángeles and Boguñá, Marián (2003). Topology of the world trade web. *Physical Review E*, **68**(1), 015101.

Servedio, Vito D.P., Caldarelli, Guido, and Buttà, Paolo (2004). Vertex intrinsic fitness: How to produce arbitrary scale-free networks. *Physical Review E*, **70**(5), 56126.

Shannon, C (1948). A mathematical theory of communication. *Bell System Technical Journals*, **27**, 379–423 623–656.

Solé, Ricard V. and Montoya, José M. (2001). Complexity and fragility in ecological networks. *Proceedings. Biological sciences / The Royal Society*, **268**(1480), 2039–45.

Squartini, Tiziano, Fagiolo, Giorgio, and Garlaschelli, Diego (2011*a*). Randomizing world trade. I. A binary network analysis. *Physical Review E*, **84**(4), 046117.

Squartini, Tiziano, Fagiolo, Giorgio, and Garlaschelli, Diego (2011*b*). Randomizing world trade. II. A weighted network analysis. *Physical Review E*, **84**(4), 046118.

Squartini, Tiziano and Garlaschelli, Diego (2011). Analytical maximum-likelihood method to detect patterns in real networks. *New Journal of Physics*, **13**, 083001.

Squartini, Tiziano, van Lelyveld, Iman, and Garlaschelli, Diego (2013). Early-warning signals of topological collapse in interbank networks. *Scientific Reports*, **3**, 3357.

Stauffer, D. and Aharony, A. (1994). *Introduction to Percolation Theory* (2 edn). Taylor & Francis, London.

Stouffer, Daniel B., Camacho, Juan, Guimerà, Roger, Ng, C.A., and Amaral, Luis A.N. (2005). Quantitative Patterns in the Structure of Model and Empirical Food Webs. *Ecology*, **86**(5), 1301–1311.

Tacchella, Andrea, Cristelli, Matthieu, Caldarelli, Guido, Gabrielli, Andrea, and Pietronero, Luciano (2012). A new metrics for countries' fitness and products' complexity. *Scientific Reports*, **2**, 723.

Tinbergen, Jan (1962). Shaping the World EconomySuggestions for an International Economic Policy. Technical report, The Twentieth Century Fund.

Tumasjan, Andranik, Sprenger, Timm O., Sandner, Philipp, G., and Welpe, Isabell M. (2010). Predicting elections with Twitter: what 140 Characters Reveal about Political Sentiment. *ICWSM*, **10**, 178–185.

Tyler, Joshua R., Wilkinson, Dennis M., and Huberman, Bernardo A. (2003). Email as spectroscopy: automated discovery of community structure within organizations. In *Communities and Technologies* (ed. M. Huysman, E. Wenger, and V. Wulf), pp. 81–96. Kluwer, B.V. Deventer, The Netherlands.

Vitali, Stefania, Glattfelder, James B., and Battiston, Stefano (2011). The network of global corporate control. *PloS one*, **6**, 1–36.

Wasserman, Stanley and Faust, Katherine (1994). *Social Network Analysis Methods and Applications.* Cambridge University Press, Cambridge.

Willis, John Christopher and Yule, George Udny (1922). Some statistics of evolution and geographical distributions in plants and animals and their significance. *Nature*, **109**, 177–179.

Yule, George Udny (1925). Mathematical theory of evolution based on the conclusions of Dr. J.C. Willis FRS. *Philosophical transactions of the Royal Society of London. Series B, Biological sciences*, **213**, 21–87.

Zlatić, Vinko, Bianconi, Ginestra, Díaz-Guilera, Albert, Garlaschelli, Diego, Rao, Francesco, and Caldarelli, Guido (2009). On the rich-club effect in dense and weighted networks. *The European Physical Journal B*, **67**(3), 271–275.

Zlatić, Vinko, Božičević, Miran, Štefančić, Hrvoje, and Domazet, Mladen (2006). Wikipedias: Collaborative web-based encyclopedias as complex networks. *Physical Review E*, **74**(1), 016115.

Index